国家自然科学基金项目“中国古城水系营建的学说及历史经验研究”资助（项目号：51278197）
教育部人文社会科学研究规划基金项目“明清广州府宗族村落空间形态格局研究”资助（项目号：14YJAZH019）

UNESCO
Asia-Pacific
Heritage
AWARDS
for
Culture Heritage Conservation

联合国教科文组织亚太文化遗产保护奖获奖工程范例

鲍家屯水碾房·兆祥黄公祠

吴庆洲　冯　江　徐好好　著

中国建筑工业出版社

图书在版编目（CIP）数据

联合国教科文组织亚太文化遗产保护奖获奖工程范例 鲍家屯水碾房·兆祥黄公祠/吴庆洲，冯江，徐好好著.—北京：中国建筑工业出版社，2015.2
ISBN 978-7-112-17872-8

Ⅰ.①联… Ⅱ.①吴… ②冯… ③徐… Ⅲ.①农田水利－修复－安顺市②祠堂－文物修复－佛山市 Ⅳ.①S27②K928.75

中国版本图书馆CIP数据核字（2015）第043204号

责任编辑：吴宇江
责任校对：张 颖 刘梦然

联合国教科文组织亚太文化遗产保护奖获奖工程范例
鲍家屯水碾房·兆祥黄公祠
吴庆洲 冯 江 徐好好 著
*
中国建筑工业出版社出版、发行（北京西郊百万庄）
各地新华书店、建筑书店经销
北京锋尚制版有限公司制版
北京方嘉彩色印刷有限责任公司印刷
*
开本：880×1230毫米 1/16 印张：14½ 字数：363千字
2015年9月第一版 2015年9月第一次印刷
定价：135.00元
ISBN 978-7-112-17872-8
（27133）

序

我很荣幸为本书作序。我认为，贵州安顺鲍家屯水碾房修复工程和广东佛山兆祥黄公祠修复工程是非常重要的遗产保护示范项目，它们获得联合国教科文组织亚太文化遗产保护奖实至名归。

从多个角度来看，这两个项目内容丰富，兼备启发性。它们不仅有助于保护实践的方法进步，同时也对当前联合国教科文组织鼓励的保护政策起到推动作用。

这其中有几点开创性的贡献。

2009 年完成的安顺水碾房保护工程，其背后的哲学思考与 2011 年联合国教科文组织在《历史城市景观保护建议》文献中提到的保护新方法核心观念一致。也就是说，水碾房工程成功地预见了遗产保护的新趋势，它的修复与当前广义视角中的保护观点紧密相连。我想着重阐述其中的两个要点。

其一，把对特定房屋的保护干预扩展到更大范围的系统保护。针对水碾房项目而言，水是其存在的根本原因，也是整个地区的生存基础。所以，单个建筑的保护也为区域文脉的保护和延续带来了机会。这并非传统意义上的独立保存，而是对遗产系统和文化景观保护的重要方法性贡献。这个项目的价值，还因为水碾房的重要性并不在于其艺术价值，而在于历史上它对水利灌溉、农业开垦、人类聚居活动的重要作用。联合国教科文组织当前推行在建筑和城市遗产保护中不单针对艺术纪念物的保护，水碾房的类型也在保护之列。

其二，水碾房修复干预后，并非要作为博物馆使用，而是要通过改善建筑物状况，维持它在本地社区中的传统生产功能。这种保护方法保持了文化遗产的原始角色和特征，还使得本地年轻一代认识到保持传统碾磨技艺这种非物质文化传统的价值。水碾房的延续变成了日常生产活动的自然结果。这是当代遗产保护中非常重要的经验：如何引导社区积极地应对本地遗产保护。

同样的讨论也适用于佛山兆祥黄公祠的项目。无需赘述在佛山这样一个城市的中心地段保留历史建筑地标有多么重要——我也知道近年来佛山经历的大规模城市拆迁和重建，给遗产保护带来了巨大的挑战。对一个急剧丢失重要历史痕迹的城市而言，这是个令人振奋的结果。同时，用修复的祠堂容纳了粤剧博物馆的文化活动，这个改变既保留住某些原初的遗产，也把这里变成一个活跃的、有价值的地方，吸引大众关注。无论作为礼仪建筑还是展览建筑，兆祥黄公祠的保护能发生在一个工业化大城市的中心地点都是难能可贵的。

和水碾房修复工程一样，2005年完成的兆祥黄公祠修复，其讨论的议题也符合当前联合国教科文组织定义保护成功与否的核心判断。

本书展现的是吴庆洲教授最重要、最具前瞻性的科研和遗产保护实践项目。我强烈推荐保护工程的实践者、年轻一代的研究者、学生仔细阅读本书。这是一本思考未来发展和工作方法的重要文献。（徐好好译）

齐珂理（Paolo Ceccarelli）[1]

联合国教科文组织地方性城市和区域可持续发展规划委员会主席

意大利费拉拉大学终身教授

1. 齐珂理（Paolo Ceccarelli）教授是国际知名的城市规划、城市遗产保护及发展研究专家。意大利费拉拉大学城市分析和城市规划专业终身教授，城市区域和环境研究中心（CRUTA－Center for Urban Regional and Environment Research）主任，ILAUD国际建筑和城市设计实验室（International Laboratory of Architecture and Urban Design）主席。曾任威尼斯建筑学院院长，费拉拉大学建筑学院院长，美国麻省理工学院、加利福尼亚大学伯克利分校、圣克鲁兹大学客座教授，哈佛大学、早稻田大学访问学者。从20世纪60年代开始在意大利、法国、美国、古巴、埃塞俄比亚、利比亚、巴西、智利、阿根廷、中国、越南以及其他国家从事城市历史中心和遗产保护及发展项目。近期主要工作为耶路撒冷老城保护研究，印度德里老城保护及发展研究，中国广州、佛山高明等地城市保护和城市发展研究。

前言

本书呈现给读者的，是两个获得联合国教科文组织亚太文化遗产保护奖（UNESCO Asian-Pacific Awards for Cultural Heritage Conservation）的工程项目，分别是2011年获得卓越奖的贵州鲍家屯水碾房修复工程和2005年获得荣誉奖的佛山兆祥黄公祠修复工程。我与团队成员们共同完成了研究与修复设计的部分，和几乎所有遗产保护工程一样，在整个过程中与很多机构和个人进行了密切而有益的讨论和协作。

在没有获得亚太文化遗产保护奖之前，这两个建筑很少有人知晓，保护级别也都不高，兆祥黄公祠是佛山市文物保护单位，鲍家屯水碾房甚至都没有被列为文物。按照我们所习惯的标准，两项工程都是投入不高，技术和工艺也不是特别复杂的古建筑修缮，但是在国际上却获得了一定的认可，这引起了很多人的兴趣，因为我国一些著名的历史村镇和传统建筑的修缮、活化利用反而未能获奖。获奖当然不是历史文化保护的目的，但是这一现象却引人深思，本书旨在通过对两个获奖项目的讨论，来促进我国的文化遗产保护事业，推动中华传统文化和工程技术的继承与发展。

UNESCO亚太文化遗产保护奖由联合国教科文组织曼谷办事处设立，从2000年开始颁发，最初设定的三种奖励是Most Excellent Project（最卓越项目）、Excellent Project（卓越项目）、Outstanding Project（杰出项目），2001年奖项名称改为Awards of Excellence（卓越奖）、Award of Distinction（杰出奖）和Award of Merit（优秀奖），2002年开始增设Honorable Mention（荣誉奖或荣誉提名），2005年开始增设Award for New Design in Heritage Contexts（遗产环境中的新设计奖，最初以the Jury Commendation for Innovation“评委会创新奖”为人熟知），其中卓越奖为最高奖 。[1]

联合国教科文组织设立本奖项的宗旨，是力图通过获奖工程范例来提高亚太地区包括建筑遗产在内的各种文化资源的管理水平。这些文化资源不仅构成了人类对过去的集体文化记忆，同时也是未来社会建设的文化基础。联合国教科文组织特别鼓励私人和私有机构在本地区文化遗产的保护中发挥重要的作用。[2]

到2014年底为止，联合国教科文组织亚太文化遗产保护奖一共收到了来自24个国家的265处申请，其中174个项目获得遗产保护范例类奖项，9项获得遗产环境中的新设计奖项。来自中国的遗产保护项目有着不俗的表现，一共获得了3项卓越奖，4项优异奖，18项优秀奖和19项荣誉奖。我所在的华南理工大学建筑学院先后获得过2次卓越奖[3]和1次荣誉奖，与广东地区对传统的重视以及民间力量在遗产保护中的活跃程度不无关系。

1. 资料来源：http://www.unescobkk.org。括号中的翻译为笔者所加。

2. 吴运江 . UNESCO亚太文化遗产保护奖综述及我院获卓越奖评价 [N]. 南方建筑，2013(1)：P88-91.

3. 陆元鼎教授主持的从化广裕祠修复工程获得了2003年度的卓越奖。

鲍家屯位于贵州省的安顺，村落始建于明洪武年间。“碾房听音”是鲍家屯的二十四景之一，村内原有7座水碾房，只有3座留存至今，本次修复的是村民鲍奇登家的水碾房，位于门前坝的旁边，建筑具有浓厚的地方特色，屋顶用鱼鳞石板瓦，部分墙身采用水层板，结构为木制穿斗。在维修鲍家屯水碾房时，意见不一，我提出保护应采用当地的材料、传统木作和石作的技术与工艺来进行维修，保持好明代古建筑原有的风貌，而不应采取导致技术传统和文化景观断裂的工程做法。鲍世行等合作者均同意该意见，并按此进行水碾房维修。经过当地工匠的精心修复，鲍家屯水碾房得以重新为当地的村民服务，利用水力进行粮食加工，原有的功能、文化景观和生活方式得以保存。该项目获得了国际遗产保护界的高度赞誉，2011年获UNESCO亚太文化遗产保护奖的最高奖——卓越奖。2012年9月，我受邀参加在克罗地亚杜布罗夫尼克召开的联合国教科文组织“2012最佳遗产保护”会议，介绍了贵州鲍家屯水碾房维修保护项目，受到与会各国专家的高度赞赏，被认为是第三世界保护文化遗产的典范，鲍家屯水碾房项目也因此成为“国际最佳遗产保护精英俱乐部”的一员，这是我国第一个入选该俱乐部的项目。

兆祥黄公祠是佛山老城中的一座规模较大且形制非常完整的祠堂，采用了空斗墙、冷巷、风兜等创造良好微气候的技术，以及水磨青砖墙、木雕、石雕、灰塑、彩绘等工艺。整个修缮过程有多方参与，既有政府部门、科研机构和文化遗产保护的专家，也有黄祥华先生的后人，看守祠堂多年的文化热心人士和未来的使用者。因为祠堂遭受了多次破坏，相关的文献资料也较为缺乏，我带领学生们首先进行了扎实的历史研究和细致的现场调研，提出尊重原真性的修缮理念，使用传统的手工技能、材料与构造，后加处理均为可逆性的工程技术措施。修复完成以后作为广东粤剧博物馆，发挥无形文化遗产的作用，已经成为粤剧从业者和爱好者的精神家园，是佛山新的文化地标之一。2005年兆祥黄公祠修缮工程获得UNESCO亚太文化遗产保护奖的荣誉奖，被评价为地方文化遗产保护的范例。

通过这2个修复项目的获奖，我感受到了该奖项所倡导的文化遗产保护价值观、原则和对技术理性的考虑：

（1）鼓励和重视私人及私有机构在文化遗产保护中发挥重要的作用。

（2）重视遗产对当代社会的价值，延续传统文化、景观和建筑的生命力，赋予恰当的功能，尤其是公共功能。

（3）注重对遗产的历史研究、档案记录和对图纸与文献的有效整理。

（4）重视传统智慧的挖掘和运用，鼓励传统材料、技术的使用和匠艺的传承。

（5）重视技术理性和工具理性，例如强调项目的复合性、敏感性和技术一致性。

（6）强调遗产项目的社会影响和传播价值。

正因为在这些方面与 UNESCO 亚太文化遗产保护奖的特点较为契合，鲍家屯水碾房和兆祥黄公祠先后获奖。因应以上特点，本书中的两个案例都包括了项目简要情况的介绍，设计图纸与说明性文字，申报 UNESCO 亚太文化遗产保护奖的文件，对修复建筑或修复设计的研究论文，以飨读者。

但是，无论是在施工过程中还是在事后的使用中，我们都注意到了修复设计存在着这样那样的瑕疵，对传统匠艺的理解和认识也有待加深。

本书的两位合作者冯江、徐好好都曾在我的指导下完成硕士和博士阶段的学习，现在都是华南理工大学建筑学院的教师。在以上两个项目的研究、设计和申报过程中都贡献了自己的思考与努力。很高兴能够与他们共同完成本书，也希望有更多的年轻人能加入到属于全人类的遗产保护实践中来。

谨与中国传统建筑遗产的守护者们共勉。

华南理工大学建筑学院

吴庆洲

2014 年 10 月 15 日于广州

目录

贵州安顺鲍家屯水碾房修复工程

一、安顺鲍家屯及水碾房修复工程简介

1. 基本信息

安顺，古盘州地，踞省城之上游，作滇南之门户。自割一州二县以隶南笼，东西尚三百余里，南北二百里。盘江之水界其西，三岔河之水界其北，遂为黔中之奥区焉。领二州、三县，附郭县曰普定，居干脉之脊，有山曰新坡，高广数里，为府治形胜。西五十里为镇宁州，有山曰东坡，轮囷盘郁，高三十里，州之镇山也。西百四十里为永宁州，白河之水从高而落，瀑布飞流，轰雷卷雪，极天下之奇观。东六十里安平，百二十里曰清镇，滴澄河之水界其间。自郡城而东，地势渐平，每每原田流泉灌注。

……普定五枝地，其广袤几与附郭相埒。但附郭多田，而五枝多山也。其余屯军诸寨，亦与各属相错。然天龙雄镇，甲秀环东；水母汇流，壬波衍北。

……鲍屯又属东门，在治东北四十五里。东接中所屯，南界府属西陇，西界镇宁白泥，北界镇宁定下增福寨。

——《安顺府志》

地理环境

在贵州省安顺市东北部，有一个古老的村落，因为九成以上的村民姓鲍，而称为鲍家屯，简称鲍屯。鲍家屯始建于明朝洪武二年（1369年），至今已有600余年的历史。它原来是明朝“调北征南”大军一支先锋部队的军营。根据《鲍氏家谱》记载，鲍家屯始祖鲍福宝，不仅富有军事才能，而且“素裕堪舆”。他曾“观风问俗”，选择了一块“地极壮丽，脉甚丰饶”的地方，这里区位条件好，土地广阔肥沃，四周有高山屏障，全年气候温和，境内有多处岩溶泉水，孔庆瑚在《安顺府志·序》也提到了鲍家屯“夷汉错处”的人文环境。战争结束后，鲍家屯人逐渐从屯军转为屯田，由军人变成了一般百姓。为了拦水灌溉，又溢流泄洪，鲍家屯先人在这里筑起了一道拦河低坝，采用“鱼嘴分水”的方式，向下游方向开了一条1.33km长的新河，把下游河道一分为二。顺河而下，又修建5座拦水低坝和5条引水渠，使村中的大部分田地都能得到自流灌溉。同时，6座拦河坝旁又设立了6座水碾房，利用水能来碾米磨面，进行农产品加工。

这6座水碾房经过历史变迁，尽管得以侥幸保存，但还是受到了不同程度的破坏，亟待抢救性维修。其中，离村落最近的一座，修建最早，保存最好，至今仍在使用。2008年，中国城市科学研究会首席专家鲍世行教授和中国水利水电科学研究院的专家们在当地考察中，对如何修复水碾房产生了分歧。因为当时鲍家屯和水碾房都还没有被列为文物，虽然有部分专家建议按古建筑修复原则施工，但是大部分专家都主张将原建筑拆除后重建钢筋混凝土水闸取代水碾房。鲍世行教授因此特别邀请了华南理工大学建筑学院的建筑历史专家吴庆洲教授到现场，参与讨论。吴庆洲教授在考察了当地村落和水利设施的现状后，提出这批洪武年间的建筑群极有可能在未来获评国家级文保单位，并力主由当地工匠，使用原有的建筑材料和本地技术，依据原样修复水碾房，完全保持水碾房的现有功能。这个提议得到鲍世行教授的支持。

在学者们的指导下，鲍家屯村民们自己动手，历时10个月，经过修复，使水碾房得以恢复原貌，重新展示活力。整个维修保护工程严格遵守文物古迹的修复原则，使用当地建筑材料和传统建筑工艺，消除了原有建筑中的诸多不安全因素，恢复了原有的使用功能与建筑风貌。

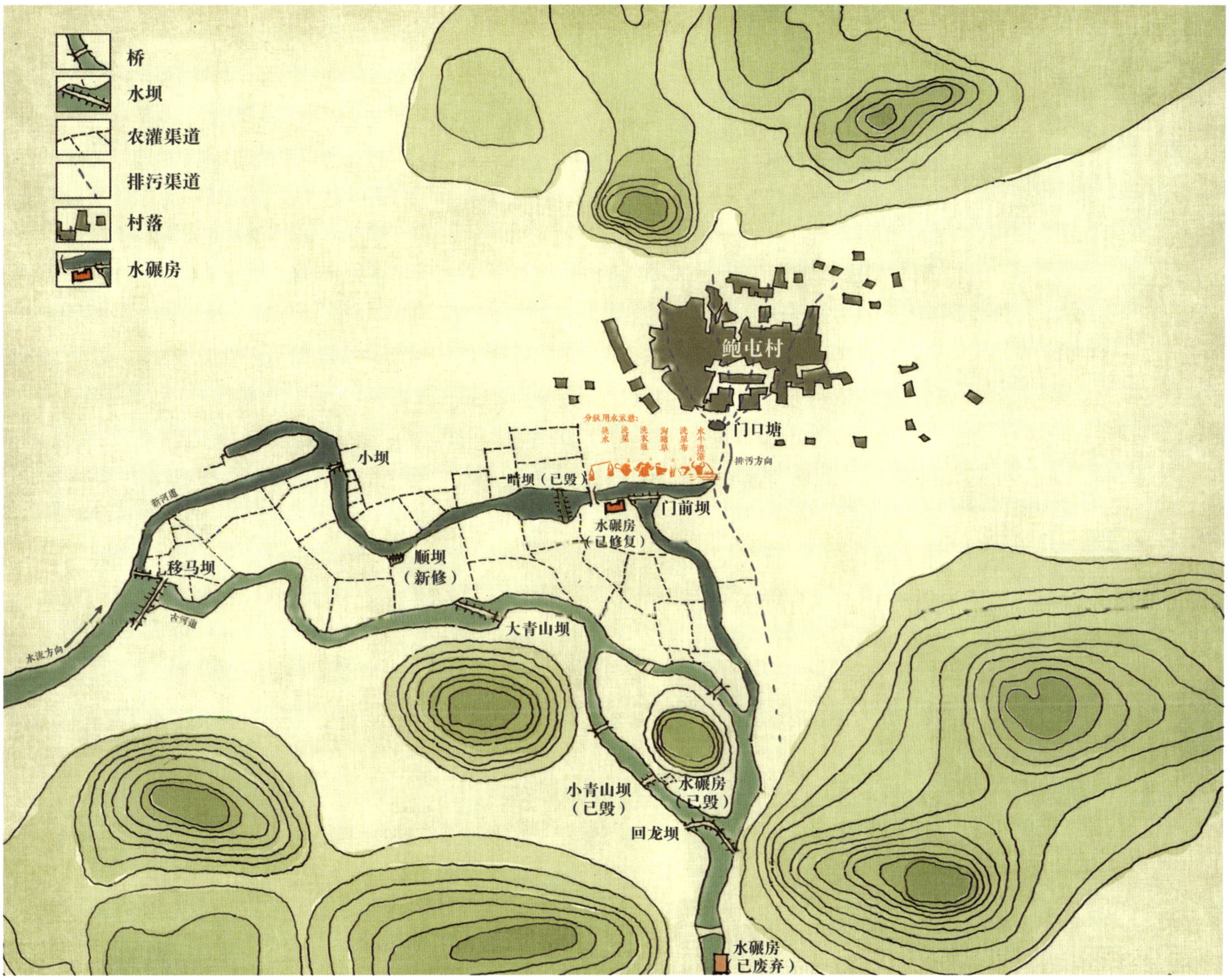

水利工程总平面图（李俊根据鲍家屯村委会《水利工程图》残稿重绘）

2. 历史研究

关于鲍家屯

明朝洪武初年，皇帝朱元璋为彻底消灭元朝梁王的残余势力，派30万大军进攻西南。为了解决庞大军队的给养问题，实行了建设水利、发展农业和驻军屯垦的政策，即所谓“三分戍守，七分屯耕，遇有紧急，朝发夕至”。当时，贵州中部以安顺为中心曾集中了数百个屯堡村落，鲍家屯是至今保存较好的一个，也是一个尘封了600余年的农村聚落活化石。

在这支军队中，有一个名叫鲍福宝的振威将军，带领家人和部下，通过深入调查研究，根据堪舆学说，选择了现在鲍家屯所在这块风水宝地，作为他们留守驻扎的地方。战争结束后，他们就在这里辟荒开田，繁衍生息，自此形成了有着悠久历史的鲍家屯。至今鲍家屯的建筑风格和村民服饰都还保留着明代的风格。

鲍福宝将军祖籍安徽歙县棠樾大和社，那里有著名的牌坊群。在带兵来西南驻扎的同时，也带来了他家乡（皖南）建设水利的经验。在村民们的共同努力下，鲍家屯逐步拥有了整套水利工程：工程充分利用当地的自然条件，建低坝提高水位，使河流分成新、旧两条，沿途又建设引蓄结合的塘坝式水利工程，并以蜿蜒曲折的渠道自流灌溉了村前3000余亩农田。工程由横坝、竖坝组成，在坝中设有高低不同的“龙口”，合理分配下游的水量，在拦河坝旁又设立水碾房，利用水能来碾米、磨面，加工农产品。当地农民骄傲地称呼这个水利工程为“黔中都江堰”。

关于水系和水利设施

鲍家屯水利设施是具有很高价值的遗产，包括了灌溉加工、生态保护和防洪排涝的重要功用。

灌溉排水：鲍家屯自流灌溉面积曾有153.3hm^2，灌区内设有引水渠，又有排水河，因此灌溉排水的效果很好，一年可以种2季，大春种植水稻，小春种小麦、油菜、蚕豆等。

水力加工：鲍家屯人利用水坝落差建造了水碾房，不仅自己的稻米用水碾，而且还承揽别处的碾米生意。

生态与环境保护：600多年来，这里的水利设施和大自然已经完全协调，营造了优美的水环境。

防洪：几百年来，尽管有2条河流流经，鲍家屯从来没有遭受过洪水灾害。

关于水碾房

鲍家屯水碾房是当地二十四景之一“碾房听音”景点中的主要建筑，位于进村必经之路的左侧，在一片田坝中间十分抢眼，是全村的标志性建筑之一。因此水碾房的修复对于展示鲍家屯的历史文化内涵和建立完整的景观体系都有重要的意义。

S 形的拦水低坝

拦水坝上的龙口

拦水坝和水碾房

3. 关于修复的思考

修复中的技术和做法

由于社会的变迁和诸多历史因素，大多数水碾房都遭到了严重的破坏，仅有一座完整地存留下来，而且还能继续靠水力来碾米，弥足珍贵。

水碾房的优点是不消耗燃料，不使用畜力，不污染环境，完全符合当前“节能、低碳、环保”的原则。其动力原理是将水的位能转变为动能，设施简单、操作方便，为当地的农产品加工提供广泛的服务，深受农民的欢迎。可以说在历史上，特别是在农耕社会中做出了不可磨灭的贡献。只是近年来由于农村广泛使用电力，水碾房的作用已经大大地降低了。鲍家屯在历史上曾经有过7座水碾房，但目前大多已经废弃，至今尚存的3座中，也仅有这一座水碾设施仍能运作。

修复项目中的水碾房位于村前水稻田中间，位置突出。建筑北面有小河流经，南侧有一小院，西北处有石桥一座。水碾房直接建于水边，当水流从北面小河往南面较低处跌落时，冲击位于建筑底部的水车，带动水碾工作。由于年久失修，在修复以前，尽管水碾还能继续工作，但该水碾房的状况不容乐观。

外观：明间（堂屋）的台阶、门和屋内的石磨保存尚好。两侧次间东西各有一扇窗保留。东侧一间，已没有木板保护，而被现代的煤灰砖填补。西立面下部较为完整，但上部已坍塌，穿斗架的木结构直接暴露出来。其他立面的外墙，石板已经破损，木板也已腐朽，缺口处被各种现代材料填塞。屋面石板瓦，除东侧一间部分倒塌以外，其余保存尚好。

结构：整个建筑的木穿斗架已经向西倾斜，成为危房。

修复工作所采用的工艺一律严格按照贵州当地传统建筑的做法，并遵守文物保护单位修复的原则：结构上利用原有的穿斗构架，消除其不安全因素，但不改变基本结构；继续其原有使用功能；就地取材，并尽可能利用旧料，不使用混凝土等现代建筑材料；采用古老的工艺，保留原材料的不规则形状；房屋的样式及装饰一律按照当地传统和原有风格来做。

维修前的水碾房

4. 修复设计

项目对结构部分的屋架（穿斗架）进行了大修。穿斗架尚完好，但屋架已倾斜，维修时进行了扶正。围护结构（墙、屋顶）进行了修补，其中墙面，不同高度用不同的做法，充分体现地方建筑的多样化。勒脚部分，为了防止洪水冲刷，用石条砌筑；窗台以下用“水层板”围护；窗间墙用木板拼接；窗上部用“竹篾墙”，用于外墙非承重的围护结构，墙面有木框架，中用用竹片进行编织，表面采用了传统工艺，在竹篾上面涂抹牛粪（这是一种特殊的工艺。竹篾墙，为了坚固起见，一般表面都涂抹石灰浆或石灰黄泥砂浆；这里的砂浆加了牛粪，是为了洪水来时，墙皮能很快脱落，室内的洪水能很快排出）。西立面的墙体位于河流来水方面，为抵抗洪水冲击，采用“平穿墙”做法（见后述平穿墙做法）。东立面按原状修复。屋顶用古老的工艺，仍用石板瓦（石板瓦屋面，当地有两种做法，早期石板不规则，铺砌比较自由，晚期石板较规则，铺砌呈鱼鳞状。水碾房采用早期的、古老的工艺做法）。水利设施中原来的水车及水磨均比较完好，仍保持原样。其他装饰（门、窗等装饰）中缺损的窗花参照保留的西侧次间窗花重做。

保护项目中的修复方法坚持安全的原则，做到结构上安全、合理；采用传统工艺，保持地方建筑风貌；材料尽量用原有建筑上拆换的材料；坚持可持续发展的原则，维修后仍可继续使用、展示。因为水碾房设计功能十分科学合理，因此修复时没有对原有功能作任何改变。

建筑就地取材，采取了石板瓦、水层板等当地常用传统建筑材料。修复时腐朽、断裂的木料从安全要求出发进行替换，但尽可能利用旧料，以保持原有风貌。旧料先清除灰尘，用砂纸打磨，最后用桐油刷底防腐。

石料也充分利用旧料。石板瓦用于屋顶，采用不规则形状，做法用古老的工艺。石料用于墙身和基础。黏结材料，用石灰砂浆；非承重部分用干砌，不用水泥等现代建筑材料。

水层板、石板瓦、石墙的砌筑工艺分别如下：

· 水层板：当地片麻岩天然形成一定厚度的质量较好的石片，用该石片竖放作室内的墙体隔断和室外窗台下的挡板。

· 鱼鳞石板瓦：用于屋顶，采用古老的工艺，天然石板形状不规则，铺砌比较自由。

· 平穿墙：竹篾墙外的石砌矮墙，高度平屋檐的穿枋，起到围护屋身和抗击洪水的作用。

水碾房内部则贴近生活，保留着木门窗和砖木雕刻。缺损的窗花参照保留的西侧次间窗花重做。建筑继续其原有的作为水碾房的功能，并有村民长期居住其中。

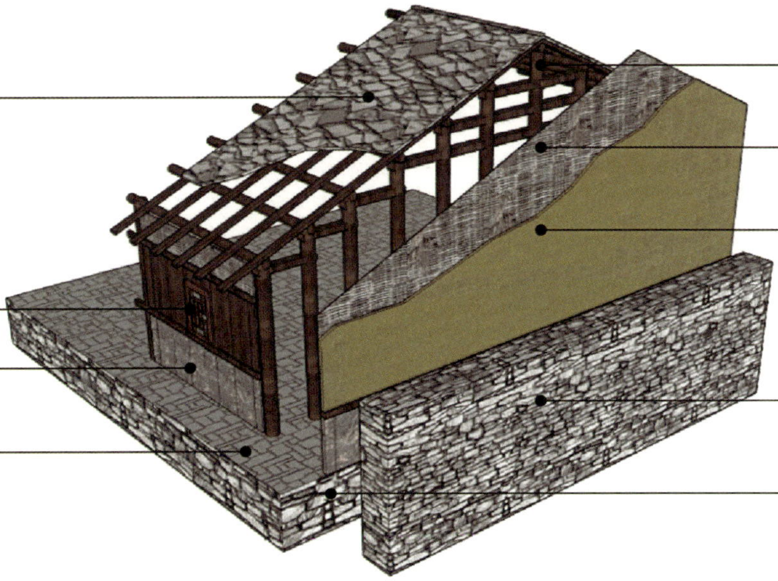

水碾房修复详图

当地常用的穿斗墙结构

当地常用的鱼鳞石板瓦屋顶

当地常用的平穿墙做法

木格窗

鱼鳞石板瓦屋顶

平穿墙和外糊牛粪的竹篾墙

竹篾墙

水层板墙面

牛粪

水碾房修复详图

5. 被扩展的修复

水碾房修复项目的目标不仅仅是消除历史遗产的不安全因素，而且是为了使历史遗产获得适当的开放性和公共性，在持续的使用中对周边环境作出积极的影响，延续传统的精神。这决定了项目关注的复合性和敏感性问题。

复合性：修复工程消除了不安全因素，恢复建筑物原真状态的同时致力于创造适当的旅游经济价值和相关的历史文化价值。此项任务贯彻于项目的全过程。

敏感性：为了满足现代社区活动的需要，计划延续水碾磨米、磨面的基本功能，在合理正常使用的前提下，提供游客参观旅游和操作的可能性。在强调了对建筑物现状的充分尊重下，遗产现状的空间布局可以基本实现这一新的要求；对于现代使用功能不可避免的新增设施（例如照明、消防等），本项目要求在降低其对于历史环境视觉冲击性的同时，保留必要的可识别性。为了不增加任何新的装饰物，少数具有展示价值的装饰物必须是可以移动的，不得固定于建筑物之上。

在水碾房修复的带动下，鲍家屯水利设施整体按时进行维修，岁修时放干河水，清理淤泥，整修水坝，保证农业用水，做到旱涝保收。2010年春节贵州地区发生大旱，可是鲍家屯仍流水潺潺，生机盎然，一片丰收景象。

水碾房的修复，不仅完整地保存了建筑的历史信息，而且恢复了建筑的历史风貌，使历史古村落焕发了新光彩。该水碾房是鲍家屯水利工程系统不可或缺的组成部分，因此水碾房的维修也展示了水利系统的完整性。目前，鲍家屯已被颁布为中国国家级历史文化名村，鲍家屯水利工程项目已申报第七批国家重点文物保护单位，获得专家一致好评，不久也将获得批准。

2011年联合国教科文组织亚太遗产保护奖卓越奖第一名授予贵州水碾房，2012年，鲍家屯项目作为亚太地区文化遗产保护的最佳范例应邀参加在克罗地亚历史文化名城杜布罗夫尼克召开的“2011世界最佳遗产范例会议”。亚太遗产保护奖评委给予鲍家屯水碾房恰如其分的高度评价：

鲍家屯古水碾房修复项目，树立了在中国进行农业景观保护的卓越范例。参与保护项目的各个机构开展了富有创见的合作，注重基地的整体性。修复项目复原了水碾房这个有着600年历史的水利设施，通过它和水坝、水道以及堤围的协作，恢复了农业生产的秩序。项目高度敏锐而小心地确保复原了水碾房的真实面貌，重现了它的历史光辉。通过对传统农业设施功能的可持续利用以及和当地文化活动的结合，这一项目为我们展现了保护在现代化发展压力下快速消失的亚洲文化景观的重要意义。

（本文根据吴庆洲、徐好好《贵州鲍家屯水碾房保护和修复研究回顾》一文改编。原文发表在2013年第1期《南方建筑》）

水碾房修复前后 1

水碾房修复前后 2

水碾房修复前后 3

鲍家屯水利工程泄洪场景

参考文献

[1] 张卫东、庞亚斌．600年鲍屯水利探考[J]．中国水利，2007（12）：51-55.

[2] 吴庆洲．贵州小都江堰——安顺鲍屯水利．南方建筑，2010（4）：78-82.

[3] 鲍世行，鲍智捷．富有特色的屯堡古村落——贵州安顺鲍屯[M]//贵州安顺鲍屯——大明屯堡古村落，2006.

[4] 扬友维、鲍中行等．大明屯堡第一屯——鲍家屯[M]．成都巴蜀书社，2008.

[5] 谭徐明、王英华、朱云枫．贵州鲍屯古代乡村水利工程研究[J]．工程研究——跨学科视野中的工程，2011，3（2）：189-195.

[6] 贵州省安顺市水利水电勘测设计院．贵州安顺市鲍屯古水利保护、恢复项目建议书[R]．2007.

[7] 彭瑛．文化生存范式下文化遗产的旅游开发——以贵州安顺鲍屯村为例[J]．安顺学院学报，2011，13（6）：1-3，26.

相关法规条例：

[1] 中华人民共和国文物保护法（2007）.

[2] 中华人民共和国文物保护法实施条例（2003）.

[3] 中华人民共和国历史文化名城名镇名村保护条例（2008）.

[4] 保护文物建筑及历史地段的国际宪章（威尼斯宪章）（1964）.

6. 历史地图辑要

历史地图引自《贵州府县志辑》

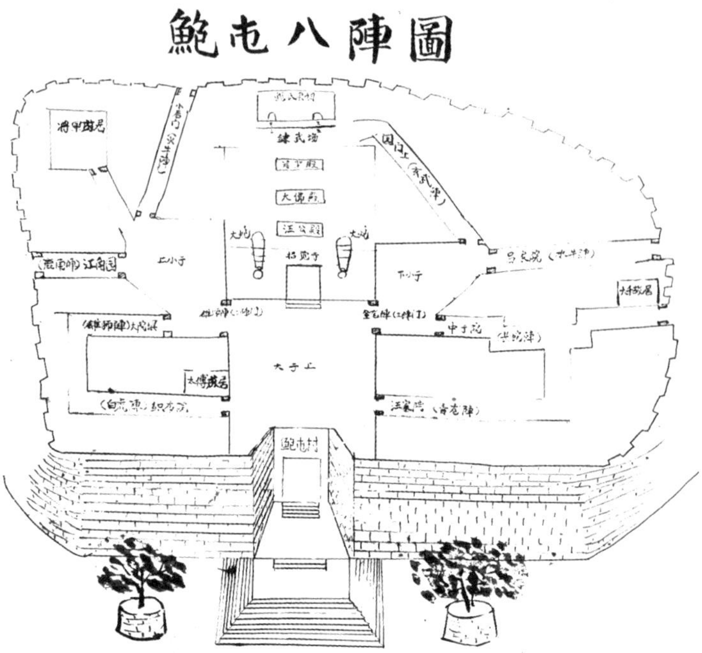

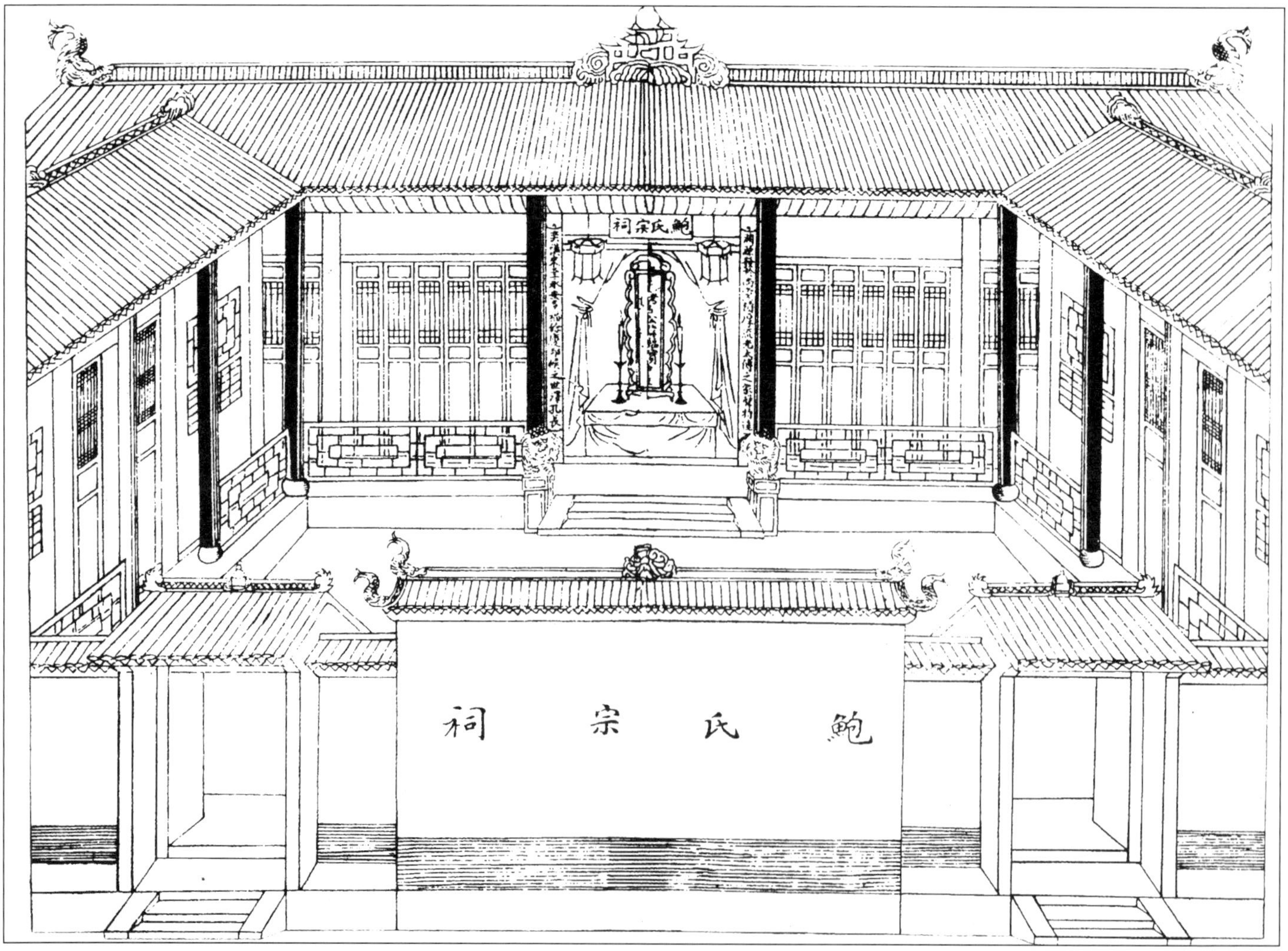

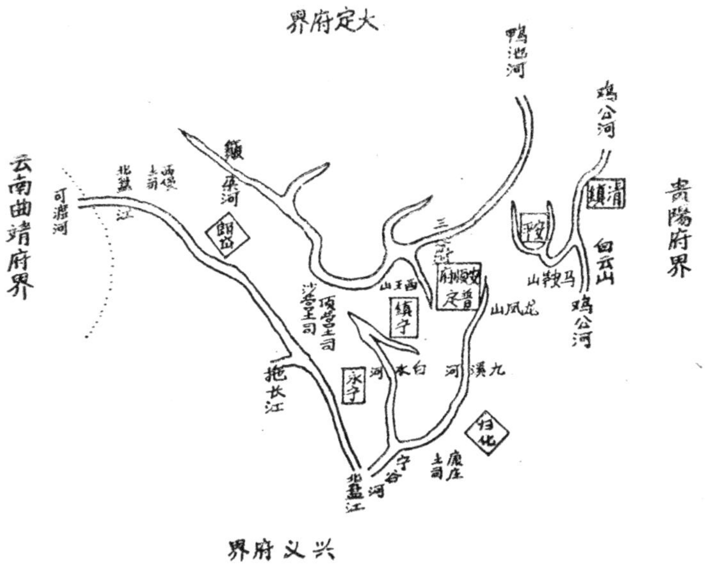

大清会典安顺府舆图

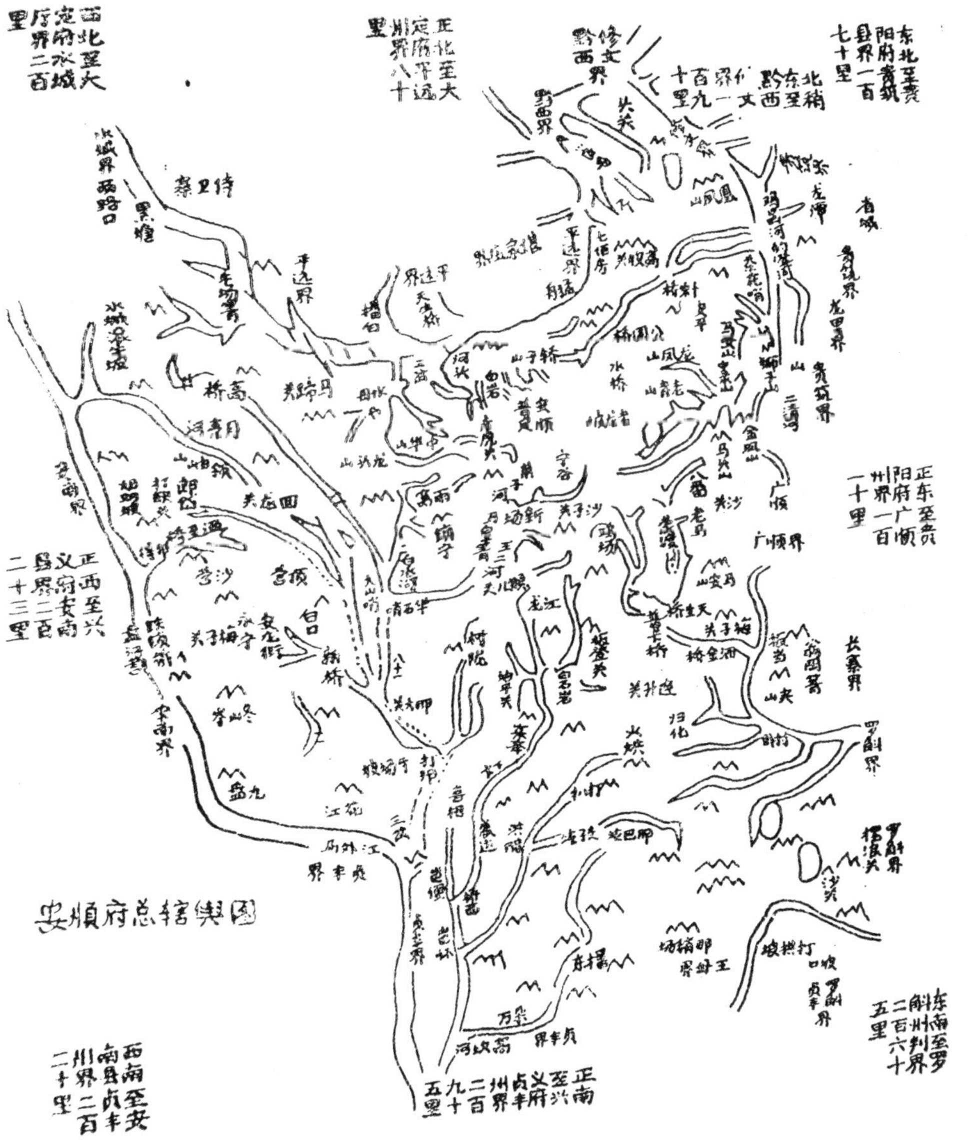

安顺府总辖舆图

西北至大定府界三百六十里
北至大定府平远州界
东北至大定府黔西州界四百里
西至南笼府安南县界一百七十里
东至贵阳府贵筑县界一百四十里
西南至南笼府安南县界四百五十里
南至南笼府永丰州界百里
东南至贵阳府广顺州与永丰州界四百里
安顺府
普定县
贵西道
安平县
清镇县
镇宁州
永宁州
归化营
郎岱营
开口营
鸭池河
白水河
盘江

安顺府舆图

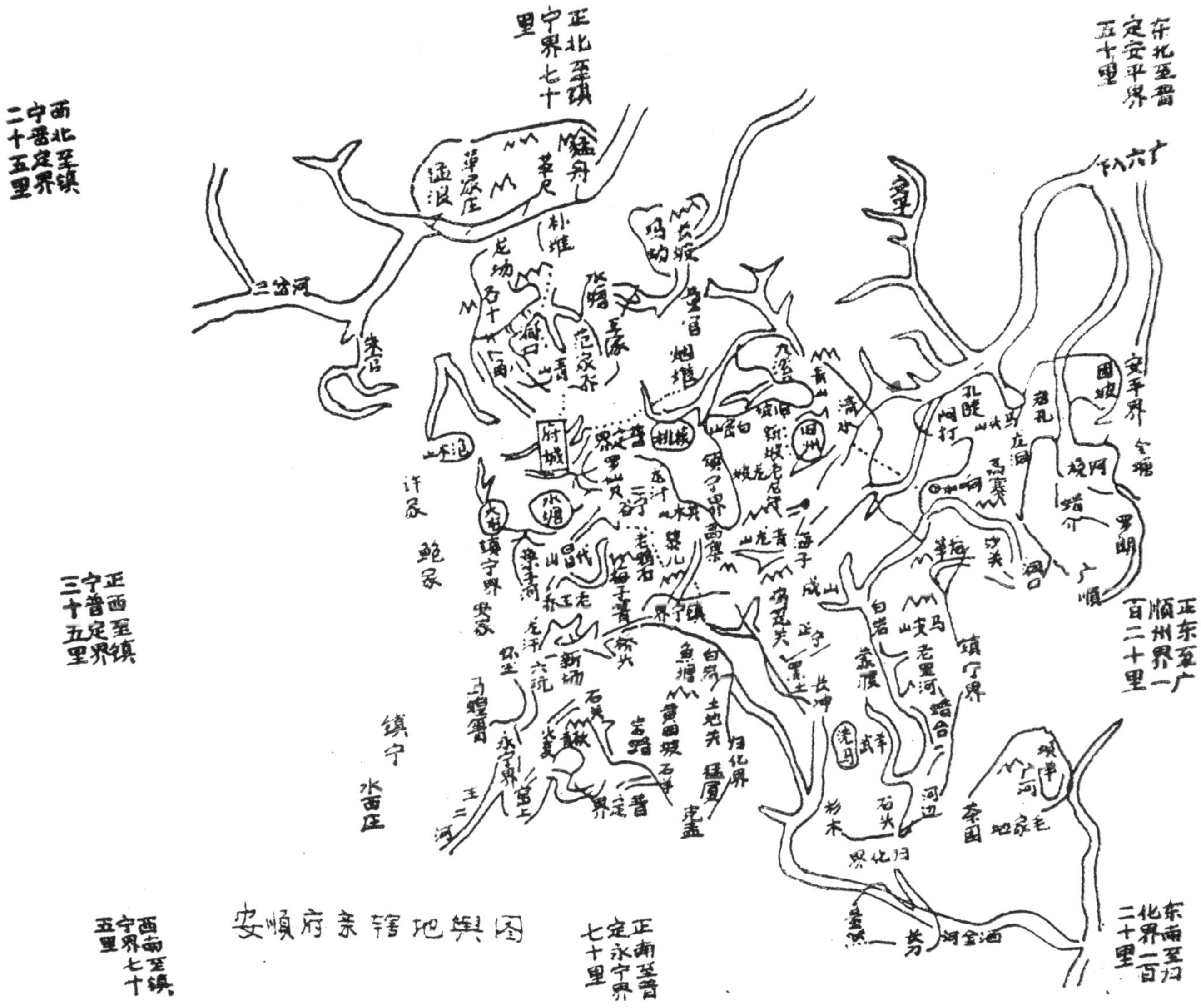
正北至镇宁界七十里
东北至普定安平界五十里
西北至镇宁普定界二十五里
正西至镇宁普定界三十五里
正东至广顺州界一百二十里
西南至镇宁界七十五里
正南至普定永宁界七十里
东南至归化界一百二十里
安顺府亲辖地舆图
府城

安顺府亲辖地舆图

二、安顺鲍家屯的水利工程和水碾房修复设计研究

1. 水利工程

都江堰名满天下，具有供水、防洪、灌溉、运输等多种功用，使川西平原从其建立之后"'水旱从人，不知饥馑'，'时无荒年，天下谓之天府'也"（《华阳国志·蜀志》）。都江堰作为世界上唯一保存至今造福人类的大型无坝引水工程，是中华民族智慧和创造力的象征，是中华民族的骄傲。2000年，"青城山——都江堰"批准列入《世界遗产名录》。

本文论述的"小都江堰"，规模虽小，却同样造福当地百姓，麻雀虽小，五脏俱全，其构思巧妙，布局合理，水资源得到综合利用，使鲍家屯600多年地杰人灵，人寿年丰。

为了对这"小都江堰"的来龙去脉有个全面的了解，还有必要对安顺鲍家屯作一个简单的介绍。

历史简况

明朝洪武初年，朱元璋为统一天下，控制西南边陲，发动了史称"太祖平滇"的战争，决心将踞守在云南一直未肯屈服的元朝梁王剌瓦尔密一举扫平。大军采取攻守兼施的策略，每攻克一处，就派士兵驻守。一边操练，一边农作，巩固后方，同时提供军粮。屯田戍边的军队就地安家。贵州在明代是通往云南和沟通川、湖、两粤的战略要地，设有许多卫所，运粮又十分艰难，因而军屯特别发达。据明嘉靖《贵州通志》及相关史料统计，明初屯驻贵州的军人在20万左右，占全国军人总数的1/10。屯堡聚落集中在贵州中部及西南部，而安顺地区尤为密集。[1]

鲍家屯村寨的"八阵图"布局别有特色。村口矗立2棵"风水古柏"，为"双柏迎宾"之景。之后是威严的屯门，然后沿中轴线由南向北依次为瓮城、汪公殿、大佛殿、关圣殿、练武场、鲍氏宗祠等建筑。中轴线的两侧，由数百座石头房屋通过8条弯弯曲曲的小巷连接起来，组成一个攻守兼备的"八卦迷魂阵"，分别命名为：青龙阵、白虎阵、雄狮阵、长蛇阵、火牛阵、金鱼阵、鹿角阵、玄武阵。传说因具备完整的防御体系，鲍家屯曾作为南明小朝廷的"临时首都"，南明永历帝朱由榔在此遣将与清军作战达2个月之久。[2]

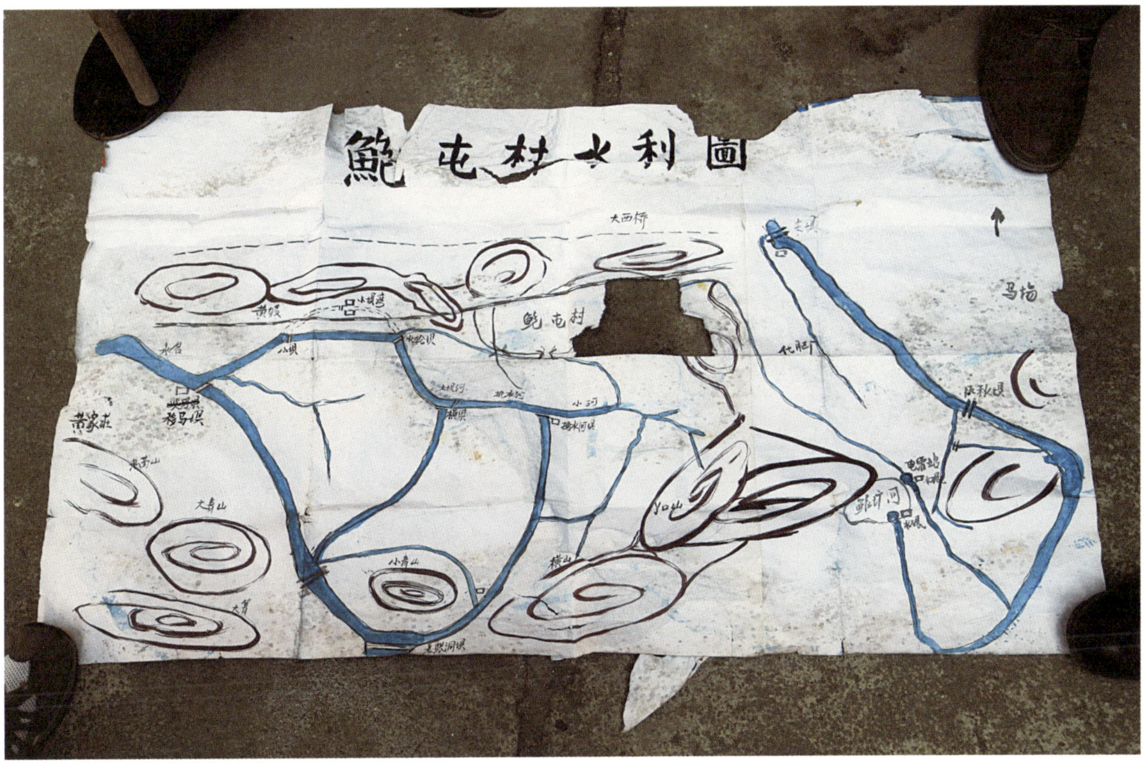

鲍家屯大门外双柏迎宾

水利建设

皖南鲍氏富有建设水利的经验，据明嘉靖三十年（1551年）程尚宽撰《新安名族志》称：东晋咸和二年（327年），鲍弘任新安太守，创兴水利，以资灌溉。他创建的鲍南堨是徽州建造最早，使用时间最长，规模最大的水利工程，1600多年来，给当地人民创造了无穷的财富。

鲍南堨位于徽州歙县岩寺镇丰乐河上。坝长约530m，干渠由岩寺潜虬山下入口，流至路口分为南北2条支渠，灌田3700余亩。[3]

歙县棠樾鲍氏还有建村结合理水的经验。元明之际，棠樾村民进行了较大规模的水系改造活动。棠樾之水发源于灵山。元至正年间（1264~1294年），鲍伯源带领族人在此河上距灵山5里之处，截流修筑大母堨，灌溉良田600余亩，确保以农耕为主的棠樾村能旱涝保收，同时引水入村，沿村南环绕如带。又引横路塘水绕村东，两股水合流至七星墩水口。[4]

始祖鲍福宝当年带来了祖籍皖南建设水利的经验，数百年来村民又不断进行水利工程建设，形成了完整的水利工程设施。整个工程系统布局合理，功能完备，设施简洁，除灌溉外，还具有供水、排泄、水力利用等功能，使鲍家屯具有便利的农业、生活用水和粮食加工等条件。

工程充分利用当地的自然条件，把大坝河分成2条河流，以及蜿蜒曲折的渠道，建成了引蓄结合的塘坝式水利工程。工程由横坝、竖坝和高低龙口组成，使不同高程的耕地均能得到充分的自流灌溉，共可浇灌3000余亩农田，使之成为旱涝保收、稳产高产的良田。人称："小菁、大菁和垭口（指山上植被良好）；横坝、竖坝顺龙潭（指充分利用水力，最后水流入九溪河）。"[5]

工程规模和设计

鲍家屯村全村土地面积3km^2，全村共有10个村民组，620户，2700人。

鲍家屯古水利设施建在型江河上流，型江河从鲍家屯村前经过。型江河属长江流域乌江水系，发源于安顺市西秀区七眼桥洞口岩上，河源高程1406m，由西向东流经郑家屯、七眼桥、三铺、鲍家屯、马场、旧州黄腊等地，由六保（河口高程1252m）进入平堰县后称为"羊昌河"，汇入清镇市红枫湖后称"猫跳河"，流入乌江。

型江河全流域面积529.2km^2，主河道长51.8km，多年平均流量9.993m^3/s，多年平均径流量31513.8万m^3；鲍家屯古水利的龙头驿马坝（又称为移马坝）以上流域面积80km^2，主河道长30.9km，多年平均流量1.51m^3/s，多年平均径流量4764万m^3，实测枯水流量0.138m^3/s。

鲍家屯村前的田堰，耕地面积1200亩，地势西高南低，分布高程1311~1307m。600多年前，前人先在型江河"水仓"处拦腰截水，修筑起既能泄洪，又能提高水位引水灌溉的壅水坝。同时采用"鱼嘴分水"的方式，向"小坝湾"方向修一条河床蜿蜒曲折到屯门前（长1.61km），把上

龙口泄水

游的1条河，分成2条河，其中新河道长1613m，老河道1009m，既能灌溉又能泄洪，环绕鲍家屯通风向阳、肥沃的田野，最后河水在“螺丝湾”回龙坝汇合，顺着原有的河床，向东南方向流去。在鲍家屯村的河流上、田坝上，共建驿马坝等6座壅水坝，7条总长5.08km的灌溉渠道，使不同高程的3250亩水田得到充分灌溉，不仅有效抗御旱灾，排洪，保证旱涝无忧，粮食稳产丰收，而且可利用水力碾米、磨面。

鲍家屯古水利利用河水落差，层层筑坝壅水，“长藤结瓜”，不但解决了农田灌溉用水问题，还为全村人提供了生活用水和水能动力。这里的壅水坝基本是顺河流建的顺坝，裹头依山丘而设，水流冲到山丘边削能减力，平缓壅水或溢过水坝，这种因势利导、因地制宜的营造方法，是使得这些石坝不毁，至今还能利用的重要原因。

主要水利设施

1）驿马坝（移马坝）

位于主河道“水仓”处，是鲍家屯村古水利工程的龙头，为“L”形坝，驿马坝下游左河道为新河，右河为老河，坝长128m，坝高1.4m，坝顶宽1m，矩形断面，坝身稳定，坝顶高程1310.4m。

2）小坝

位于新河中段上，坝长17m，坝高1.5m，坝顶宽1.1m，矩形断面，坝身稳定。

3）门前坝

位于新河鲍家屯村前，坝长81m，坝高2m，坝顶宽1.4m。

4）大青山坝

位于老河道大青山河上，坝长60m，坝高1.6m，坝顶宽1.3m。

5）小青山坝

位于老河道小青山河上，大青山坝下游250m，坝长19m，坝高1.6m。

6）回龙坝

位于新、老河流汇口小青山下游200m主河道上，鲍家屯古水利尾部，S形布置，坝长77m，坝高2.2m，坝顶宽1.6m，坝顶高程1306.2m。

鲍家屯古水利引水龙头驿马坝以上流域面积80km^2，年平均流量1.51m^3/s，多年平均径流量4764万m^3，实测枯水流量0.138m^3/s，来水量对于鲍家屯古水利灌区用水量，绰绰有余，水量较丰富。

鲍家屯古水利由不同高程的7条支渠引水灌溉，总长4080m，设计流量0.02~0.03m^3/s，渠道宽0.3~0.8m，高0.5~1.2m，比降1/1000~1/3000（现状）。[6]

驿马坝（移马坝）

小坝

大青山坝

小青山坝

门前坝

水龙坝

顺坝

水利工程特点

据中国水利报社副总编辑张卫东总结，鲍家屯水利工程有如下几个特点：[7]

结构简单实用。鲍家屯的“坝”其实就是1~2m或2~3m高，几米到几百米长短不等的砌石堤。水大时坝顶可以溢洪，有的溢流面纵剖面为优美光滑的曲线构造；水小时水走龙口和渠道，坝顶可以走路。所谓龙口，就是砌石坝上朝天开着两三个石槽。坝的一侧连接着水渠，水渠进口段也是石槽，构造与龙口一模一样。龙口分大龙口、小龙口，用石块、水草堵住其中某些龙口，就可以很方便地控制水流去向及分水流量。坝体底部一般都留有排沙孔，高30~40cm，宽不足1m，平时用石块堵住，农闲时打开底孔，利用水力冲沙，可以省去清淤淘河的辛苦。

刻意延长坝轴线。主要的坝都建在河道比较宽的地方，如果河道不够宽阔，就把大坝修得弯弯曲曲（驿马坝长坝、门前坝、回龙坝），或是在河道中斜着布置坝轴线（驿马坝长坝），有的像平缓的S形导流堤，有的却像个伸长了的秤钩子。在今天的水利工程师们看来，这样布置似乎很不合理，其实考虑到鲍家屯的坝都是低矮的溢流坝，容易漫水溢流，不难发现古人是在有意识地延长溢流坝长度，减小坝顶宽流量，降低洪水冲垮大坝的可能性。

尽量利用岩基。为了提高抗冲能力，大青山坝、小青山坝、回龙坝（当地人称作老獐洞坝）都建在基岩上，其中大青山坝基岩出露，犹如数条鳄鱼横亘河中，坚不可摧，部分地替代了人工筑坝，小青山坝则利用天然岩石作分水鱼嘴，诚可谓巧借天然助人工。

使用当地材料。坝体使用坚硬的石头，楔形砌筑；胶结材料则用黄泥巴加石灰。这些都是当地容易得到的材料，有利于岁修维护。鲍家屯甚至规定了取土的料场。坝体不要求滴水不漏，以现代的眼光看，这样更符合生态保护要求。

鲍家屯水利工程设计科学，安全系数高，没有听到过曾发过垮坝事故。

通往水碾房的弯曲渠堤

水利工程综合功用

中国古代的城市水系有十大功用，[8]鲍家屯的水利工程也有十大功用：[9]

1）灌溉排水

鲍家屯灌区自流灌溉面积曾有153.3hm^2,下游邻村有66.7hm^2，上游邻村灌区还有133.3hm^2。灌区内没有引水渠，又有排水河，灌、排两不误，一年种两季，除水稻外，还种小麦、油菜、蚕豆等。

2）水力加工

鲍家屯人利用大坝落差建造了10座水碾，不仅自己的稻米用水碾，而且还承揽别处的碾米生意。

一条引水渠把“水仓”的水引到了村前，人们的生活用水包括饮水格外方便。池塘旁边“洗衣处”“洗菜处”两截石碑以及“牛饮水处”等约定，见证了鲍家屯人对水的依赖，也见证了他们对水的合理利用、管理与保护。

3）养殖

群众在河塘里养鸭、养鱼、养鹅，用水碾加工产生的米糠和水草养猪，成为鲍家屯经济的重要补充。

4）游泳健身

鲍家屯大坝都不高，坝前形成大大小小的水塘，这些水塘是天然的游泳场，至今仍是孩子们的乐园。村里还专门指定比较隐蔽的一处河湾为“妇女洗澡处”。

5）园林造景和旅游

开河引水到村前，是创建鲍家屯村落时规划上的需要，也是营造园林风景的一种手段。鲍家屯的一个特点是它的村落布局形成了一个“水口园林”，很明显是受到皖南古村落水口园林的影响。水口园林的组成元素除自然山水外，还有路、桥、亭、古树、碑碣、牌坊，甚至还有戏台、水街、社屋、祠堂等。[10]

6）生态与环境保护

600多年来，这里的水利设施和大自然已经完全协调，营造了优美的水环境。

7）防洪

几百年来，鲍家屯人尽管身处东西两河“夹击”之地，但没有感受过洪水灾害，几乎不知洪水为何物。

8）军事防御与交通

中堡之所以成其为屯堡，不只是因为屯田，还因为军事防御功能。在这方面，水利工程也打上了屯戍历史的烙印。在鲍家屯的防御系统中，河流是不折不扣的“护城河”；在鲍家屯的道路系统中，大坝又是四通八达的快速路。

9）促进民族团结

如今鲍家屯附近老百姓当中口口相传着这样一个故事：鲍家屯原本是从一个夜郎古村箐箕凹的基础上发展起来的。马树河水自西南而来，明初北岸为汉军屯守的鲍家屯，南岸为苗寨黄家庄，汉苗划河为界，原本互不来往。这影响到了鲍福宝兴修水利的计划，堤坝修到对方的地盘不得不停工。为了能够快速打开局面，鲍福宝便与“苗王”结拜为兄弟。从此，“驿马坝”顺利建成，他不仅连接了2个村寨，而且惠及河流两岸大片土地，成为汉、苗、布依、仡佬各族群众融合团结，共建共享水利成果的象征。

管理和维护

鲍家屯水利能600多年沿用至今，历代村规民约及管理和维护是至关重要的。

据安顺鲍氏家谱记载：鲍氏置田祭祀租石，每年收32石6升（9618斤），田大小有133块，每块有坐落地名，东南西北四至分明。分布于有水利保证的各个田坝，还有秧田搭配白田，开列清单于家谱上，“以永世守”。它的收入用于助学、赈灾，维修祠堂，建设公益事业，修河整沟，修路架桥，择老成谨慎、热心公益事业者管理财务，如有不称职者教之换之。每年春秋两季将财务公布透明，维护族产，当年收入当年用光。每年春耕生产前进行一次大规模维修，挑黄泥巴修水坝。

鲍家屯村村规民约明确规定：爱护河流、河岸、路道。河内不准乱丢生活垃圾、农业生产废弃物；河岸河坝不得损坏，岸上的自然植物不准乱砍乱折乱挖；河内严禁炸鱼、毒鱼、电鱼。如有违反，除要求恢复原状外，还要视情节按有关法律法规惩处。[11]

应中国城市科学研究会鲍世行先生的邀请，笔者于2008年和2010年2次赴鲍家屯考察，2010年3月初，正是云贵地区百年不遇的大旱之时，鲍家屯仍是青山绿水，金色的油菜花漫山遍野，景致十分迷人，一点也无大旱之象。周干峙院士，罗哲文老先生也前来考察，对鲍家屯的古水利工程及屯堡文化十分赞赏。保护好这一古代珍贵遗产，让它更好地为当代的鲍家屯人造福，这是我们共同的心愿。

（本文根据吴庆洲《贵州小都江堰——安顺鲍屯水利》一文改编。原文发表在2010年04期《南方建筑》）

鲍家屯的水口园林景观之一，小青山和永安阁

水碾房供水

渠边洗衣的妇女

军傩表演

屯堡服饰

如画美景

参考文献

[1] 魏皎，汪永平，王盈．贵州安顺屯堡聚落防御性分析[J]．华中建筑，2010（3）：157-160.

[2] 杨友维，鲍中行等．大明屯堡第一屯——鲍家屯[M]．成都：巴蜀书社，2008：15-49，121-127.

[3] 贺为才．徽州城市村镇水系营建与管理研究：236．华南理工大学，2006：236.

[4] 东南大学建筑系．徽州文物管理所编著．棠樾[M]．南京：东南大学出版社，1993：1.

[5] 鲍世行，鲍智捷．富有特色的屯堡古村落——贵州安顺鲍家屯[M]．鲍世行．谭徐明主编．贵州安顺鲍家屯——大明屯堡古村落．2006.

[6] 贵州省安顺市水利水电勘测设计院．贵州安顺市鲍家屯古水利保护、恢复项目建议书[R]．2007.

[7] 张卫东．庞亚斌．600年鲍家屯水利探考[J]．中国水利，2007（12）：51-55.

[8] 吴庆洲．中国古代城市水系[J]．华中建筑，1991（2）：55-61.

[9] 张卫东．庞亚斌．600年鲍家屯水利探考[J]．中国水利，2007（12）：51-55.

[10] 鲍世行．鲍家屯的水口园林[N]．中国水利报，2007-4-19（8）.

[11] 尹东跃．加强古水利工程保护意义深远[N]．中国水利报，2007-4-19（8）.

鲍家屯村民维修田间水利设施（吴忠贤摄）

鲍家屯古水利工程设施现状（吴忠贤摄）

2. 水碾房修复设计研究

本次修复项目的对象鲍家屯水碾房及其相关附属水利设施是鲍家屯水利系统工程的重要组成部分，经过历史变迁,尽管得以侥幸保存下来,但它的许多方面还是受到了不同程度的破坏，亟待抢救性维修。尚存的这一座保存基本完好，且水碾设施仍能运作，这是十分可宝贵的。修复该水碾房，并确保其可持续利用和效益的充分发挥，在合理使用的前提下，项目还进一步提供了旅游观光的可能性，为村民们创造适当的经济价值，也弘扬了传统文化，将在发展地方经济中发挥重要的作用

2008年，修复工程开展。经过多方努力，这座水碾房及其相关的水利设施终于得以成功修复，目前处于良好的使用状况中。

项目的阶段和时间表

保护阶段	时间安排
项目相关调研	2008.2~2010.3
历史文化建筑设施测绘	2008.3~2010.5
历史文化建筑设施修复施工	2008.9~2008.12
项目竣工验收	2009.10

项目规模和费用

项目	规模
河堤修复	2623m
改造灌溉支渠	7 条，总长 4080m
修复田间渠道	3500m
恢复壅水堰	3 座
恢复水碾房	1 座

费用来源	数量
水利水电科学研究院研究课题结存经费（整修古水碾房）	30000 元人民币

水利工程维修的费用另列。

修复原则

该水碾房设计功能十分科学合理，因此修复时没有对原有功能作任何改变。修复中：

（1）坚持安全第一的原则，做到结构上安全、合理。

（2）坚持“修旧如旧”的原则。

（3）采用传统工艺，保持地方建筑风貌。

（4）尽可能利用原有材料，替换的材料尽量用老建筑上拆换的。

（5）坚持可持续发展的原则，维修后仍可继续使用、展示。

整体修复计划

整体建筑的修复坚持修旧如旧的原则：结构上采取建筑原有的穿斗构架，消除其不安全因素，从立面看出穿斗架已经向西侧倾斜，成为危房，需在维修中将其扶正，但不改变基本结构，面向水系的竹篾墙外石砌矮墙，高度平屋檐的穿枋，起到维护屋身和抗击洪水的作用；水碾房继续其原有使用功能；本次修复工作所采用的工艺一律严格按照贵州当地传统民居的做法，就地取材，并尽可能利用旧料，新的材料也要做旧，木料要先清除灰尘，用砂纸打磨，为防腐，最后用桐油刷一遍，凡用新的石料也都要“做旧”，不使用混凝土等现代建筑材料；做法采用古老的工艺，保留原材料的不规则形状；房屋的样式及装饰一律按其原有风格来做。

工程施工一律选用当地民居传统材料，并严格按照当地传统技术施工，未在保护对象上添加任何与其原有材料或做法不同的东西。强调了技术手段的一致性，即以“复原”为原则，采用传统的材料、传统的工艺来完成所有修缮工作。采用了“技术交底在前，施工在后，样板引路”的施工程序，使之成为复原修复的样本。

修复研究中参照了关于文化遗产保护的国际文件，诸如：《奈良原真性文件》（*Nara Document on Authenticity*）、《威尼斯宪章》（*Venice Charter*）和《乡土建筑遗产宪章》（*Charter on the Built Vernacular Heritage*），以及中华人民共和国文化遗产保护相关法律，中国文物保护法和相关的设计规范等。后期使用中该水碾房将仍用作碾磨谷物，并将同时作为村民的住所使用。后期维护由相关文化保护部门管辖，由当地工匠进行定期检修。

近年来，鲍家屯附近的一些屯堡古村落，热衷于农村的“现代化”，错误地拆除村口的门楼，拆宽了村里的主要街道，以便引汽车进村，同时把古民居改建为体量较大的仿古建筑，致使古村落的结构发生了根本的改变，古村落的历史风貌荡然无存。这种错误的现象使学者扼腕叹息，深感遗憾。鲍家屯被批准为历史文化名村和鲍家屯水碾房严格复原修复，为大家树立了一个无声的样板，它的示范作用是无可比拟的。

（本文及图片部分引自徐好好《前现代建筑技术一课》。原文发表在2015年第1期《新建筑》）

选址图（李俊根据鲍家屯村委会提供手绘残稿重新绘制）

总平面

维修前（吴忠贤摄）

维修中（吴忠贤摄）

维修后（吴忠贤摄）

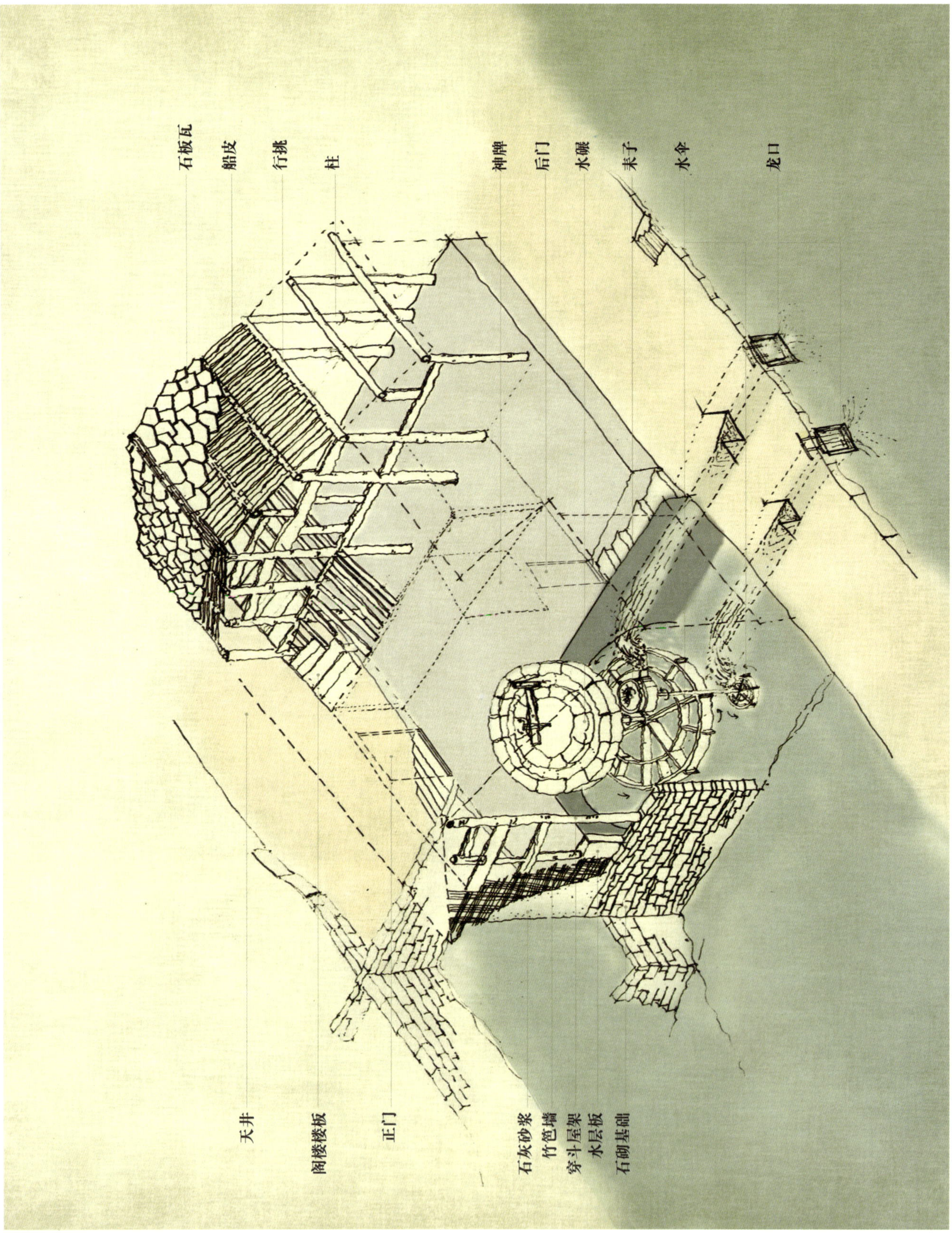

水碾工作示意轴测图

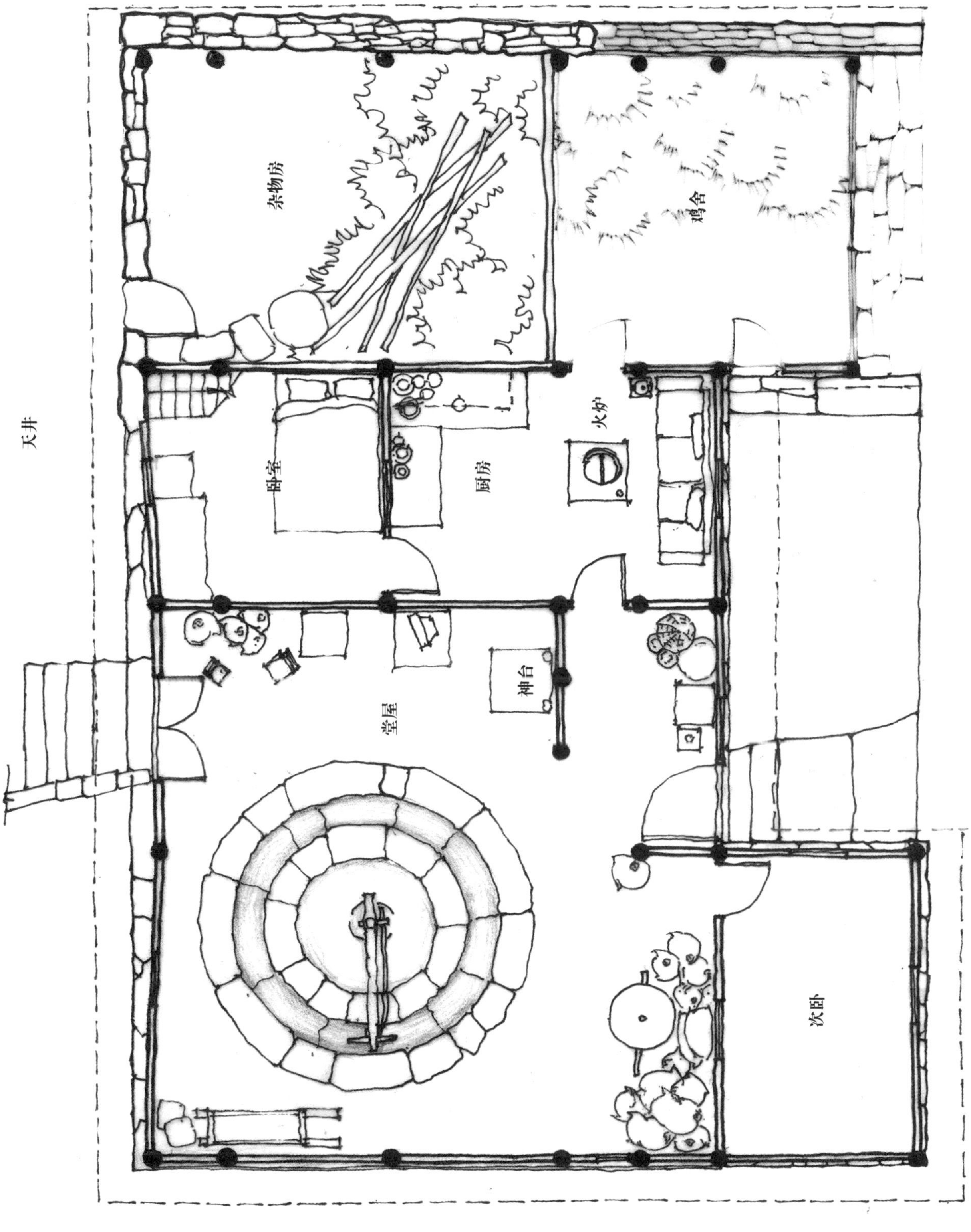

首层平面图修改

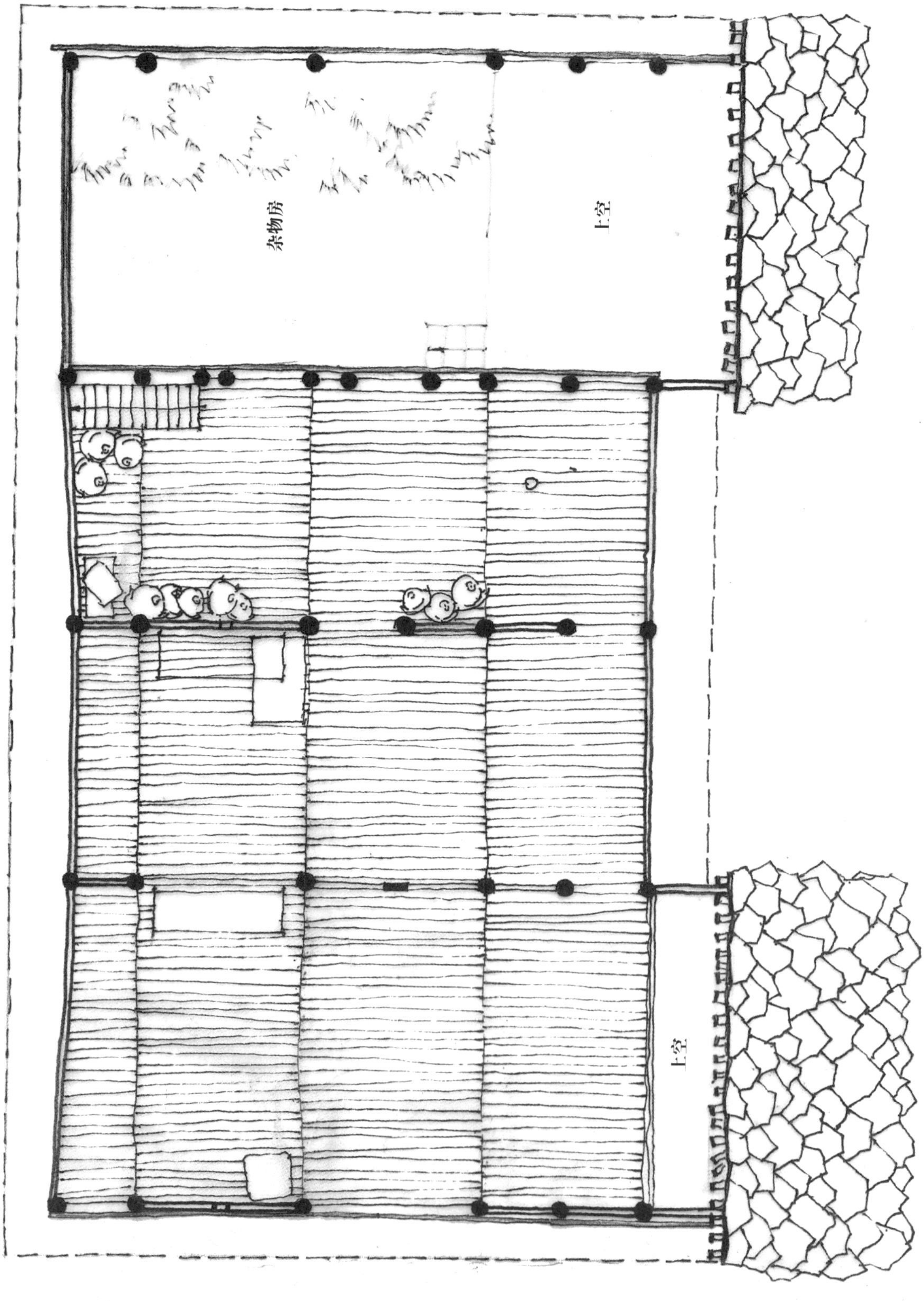

二层平面

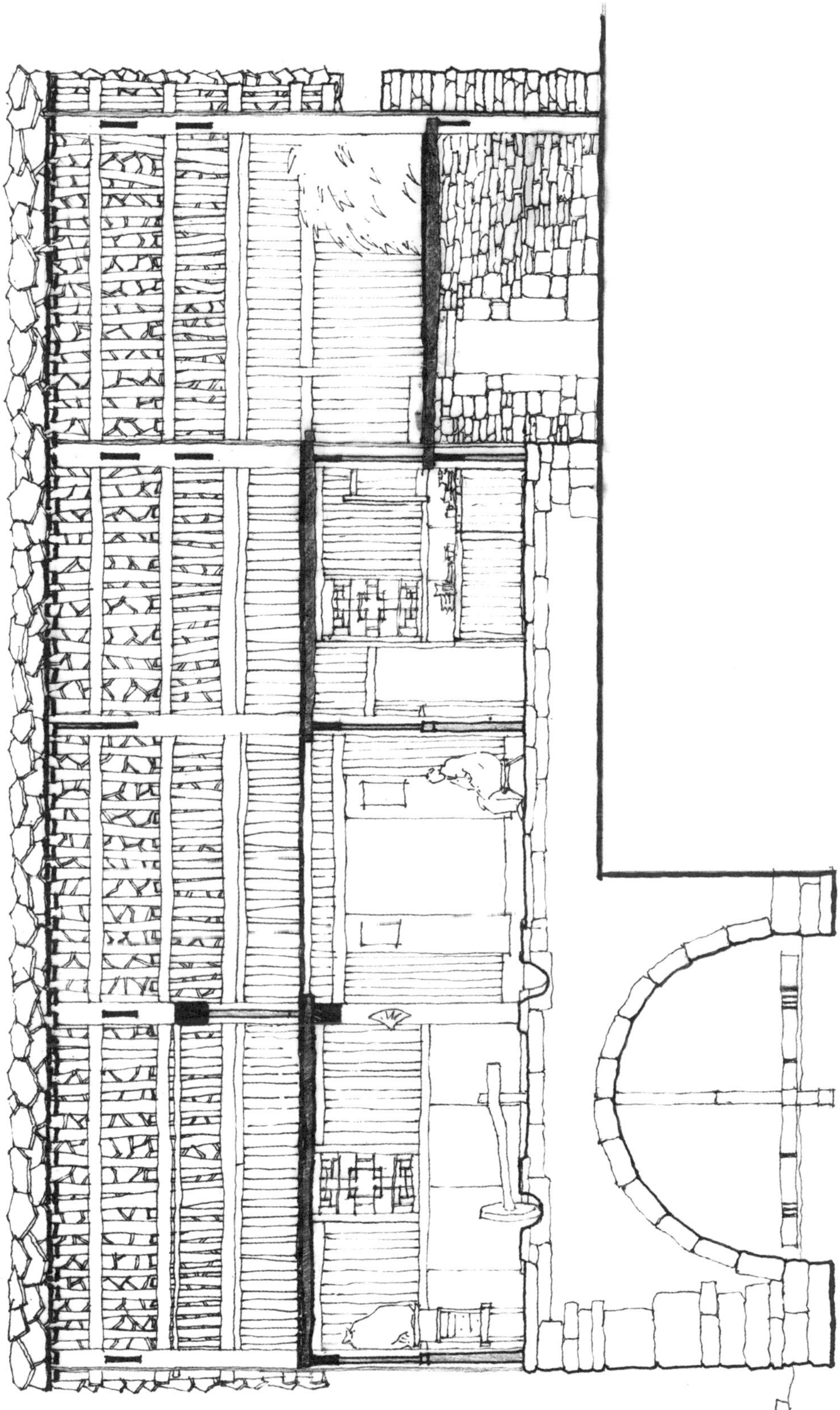

水碾房纵剖面

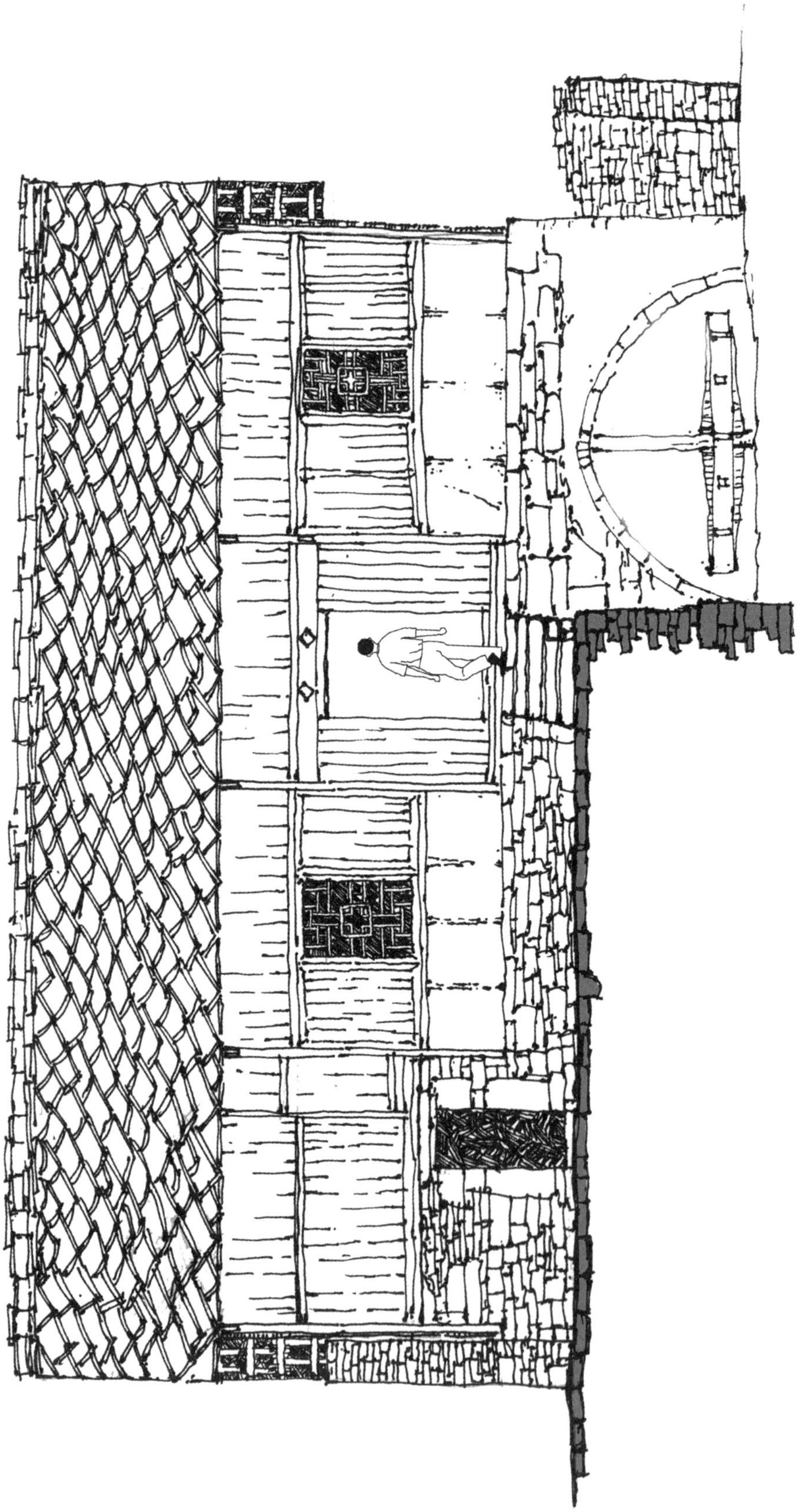

南立面

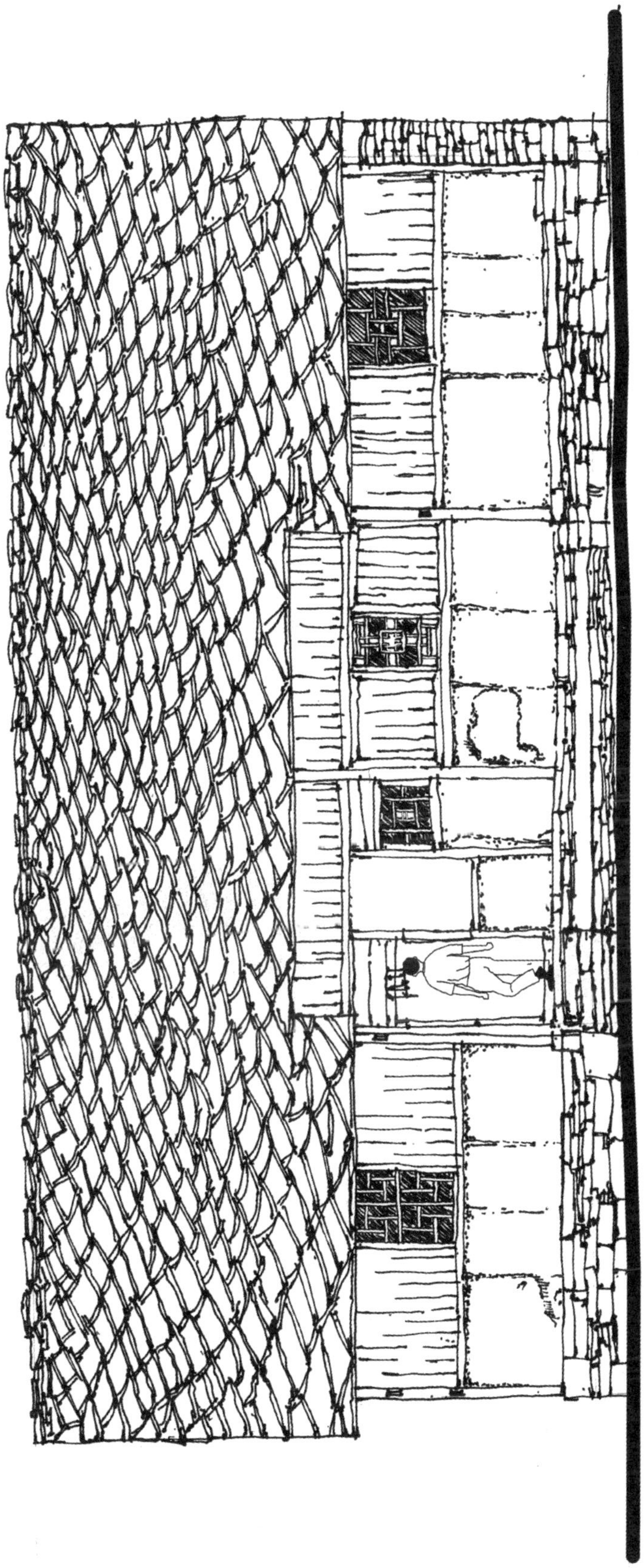

北立面

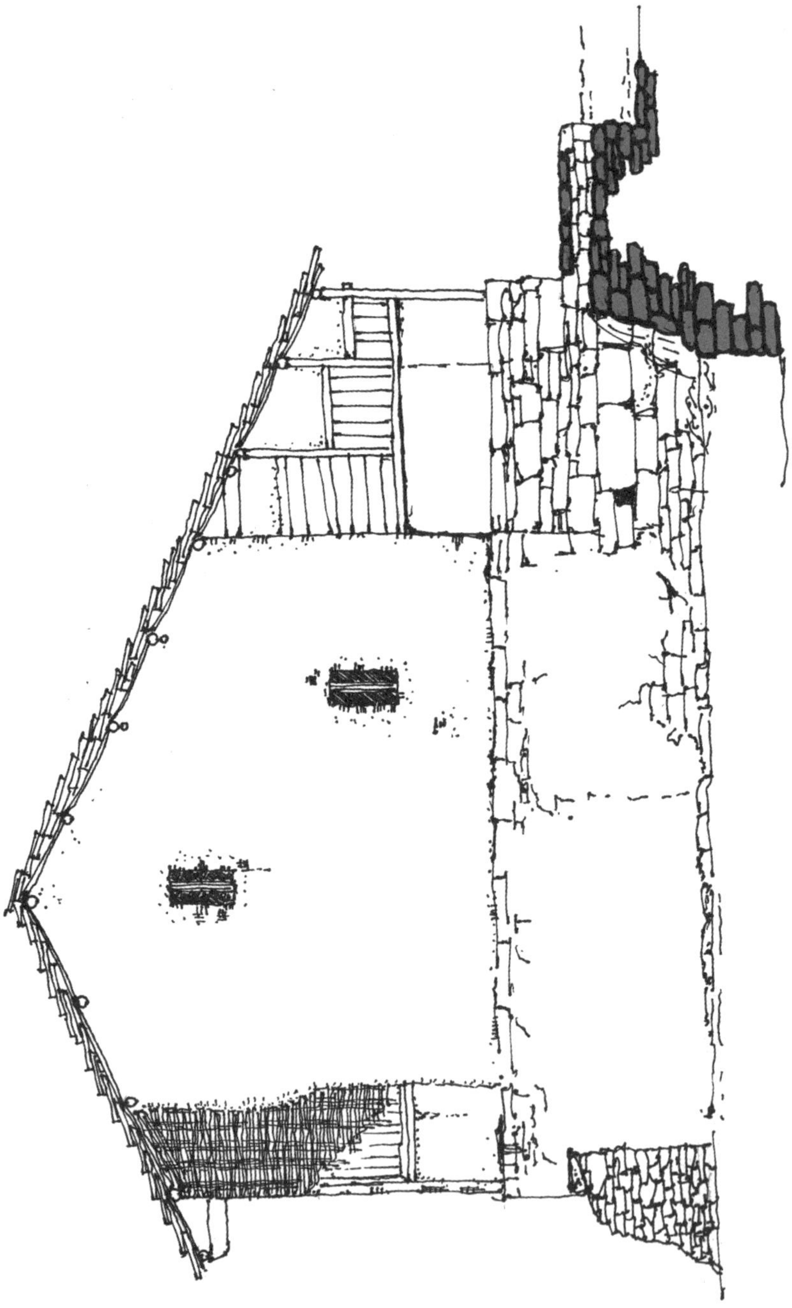

东立面

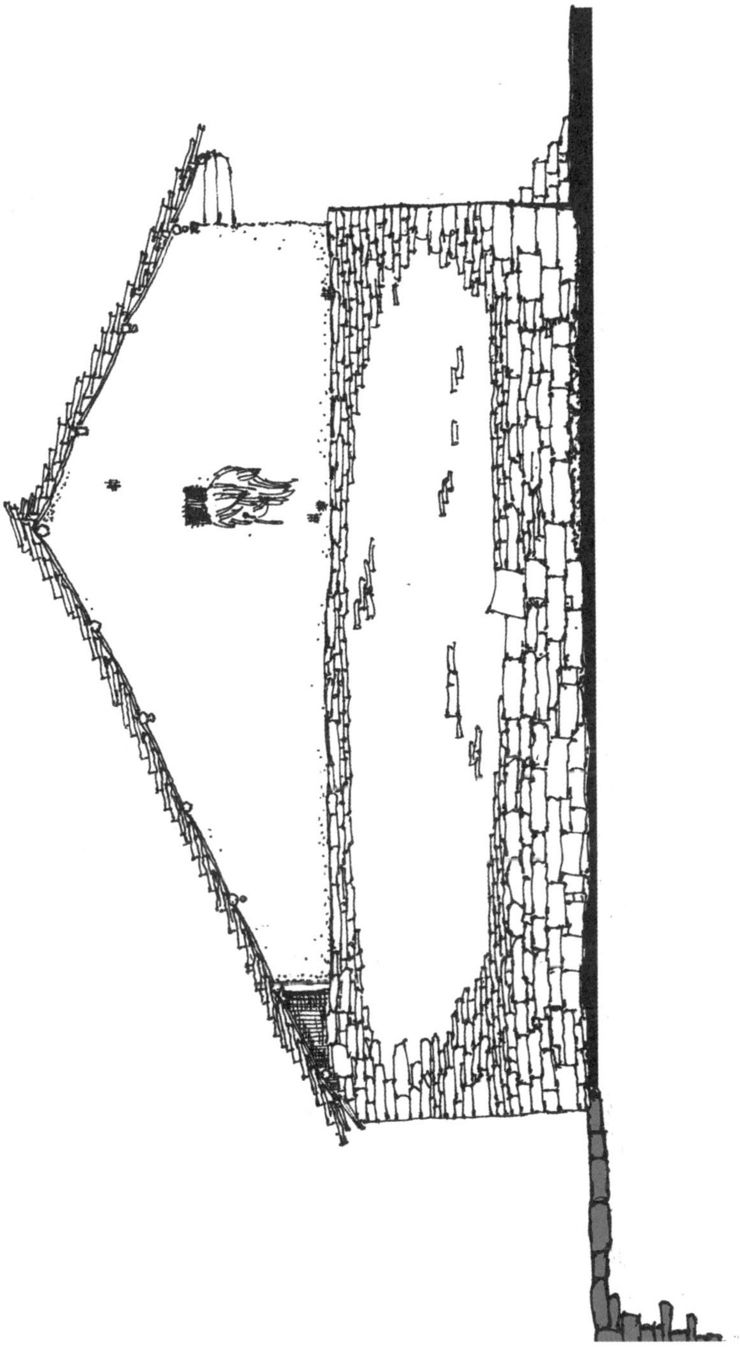

西立面

各榀架示意图

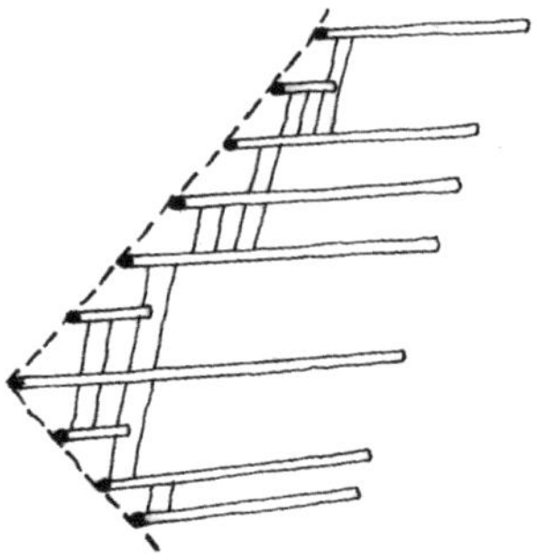

⑤

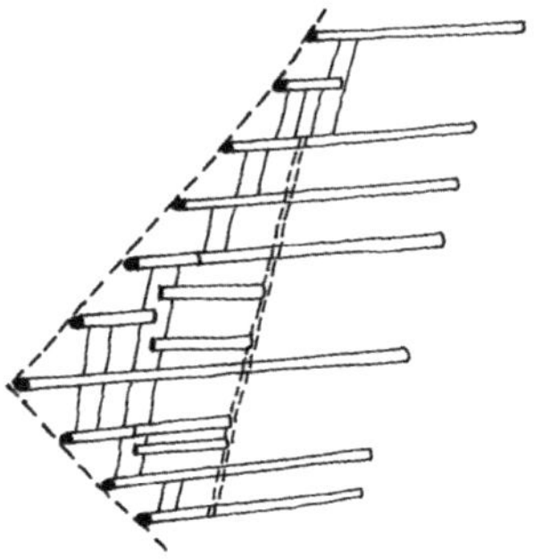

④

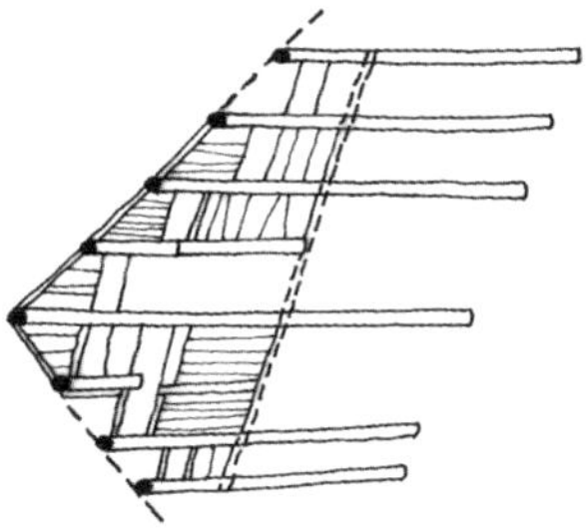

③

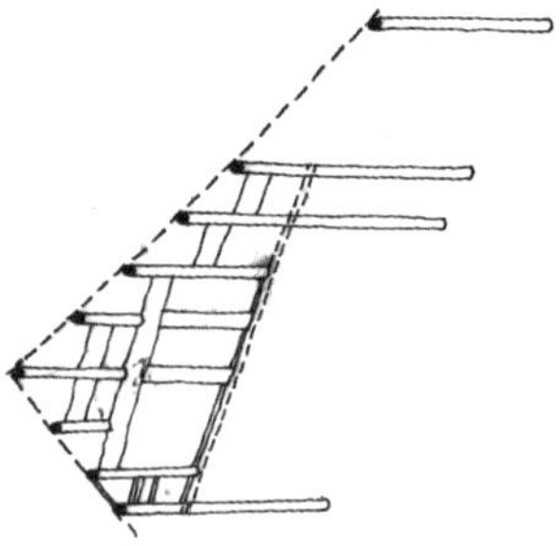

②

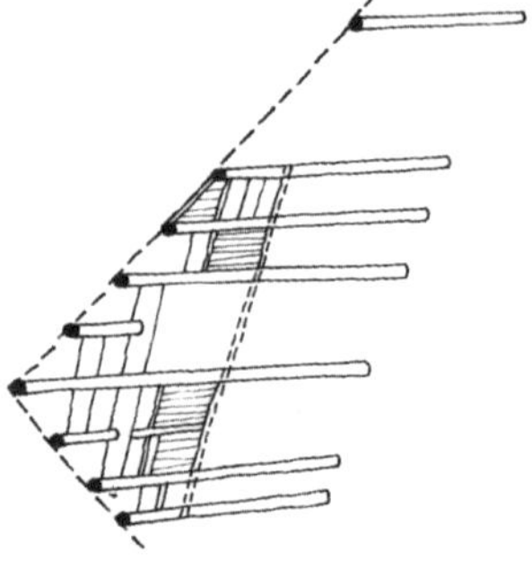

①

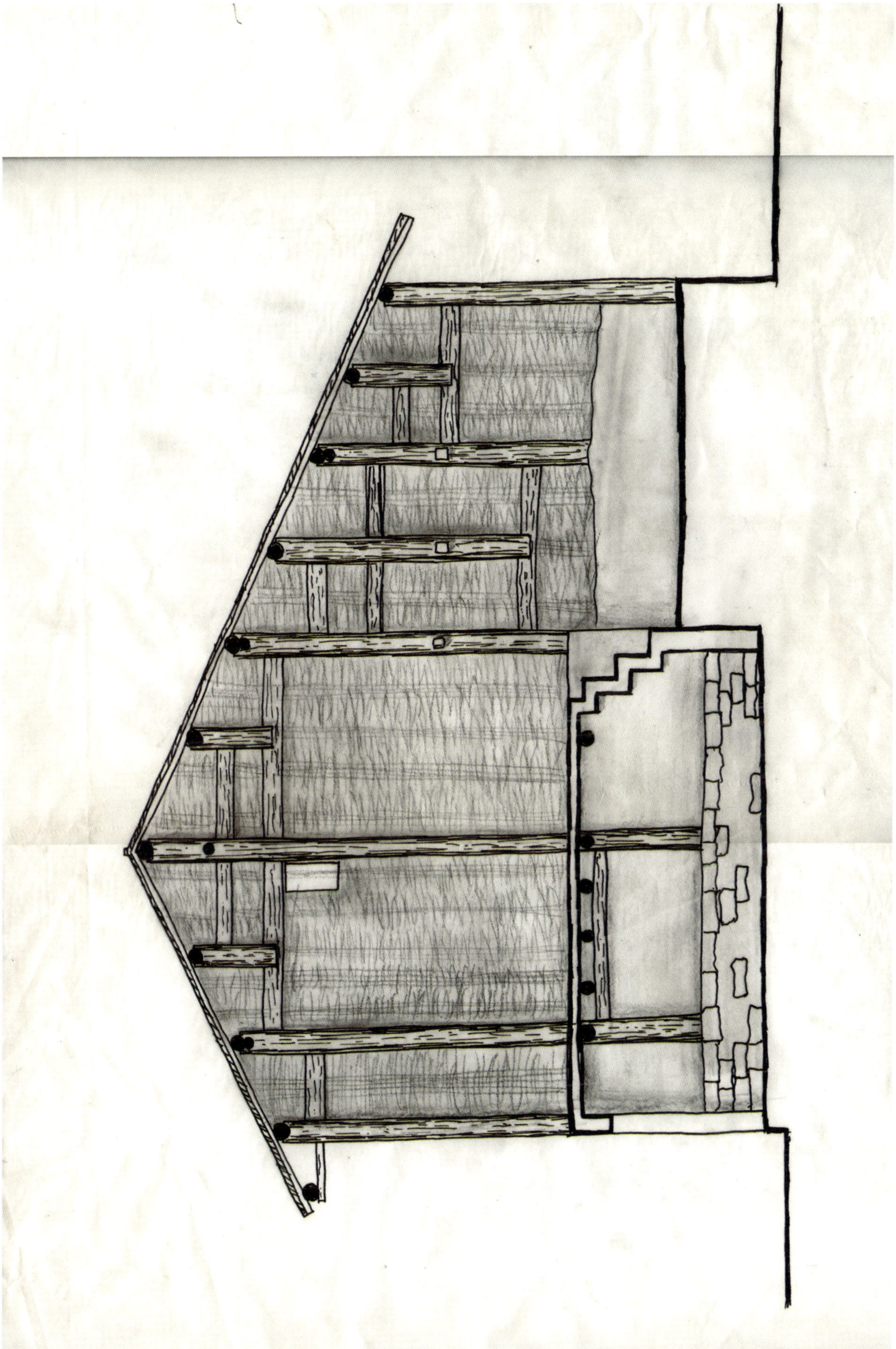

仓库剖面

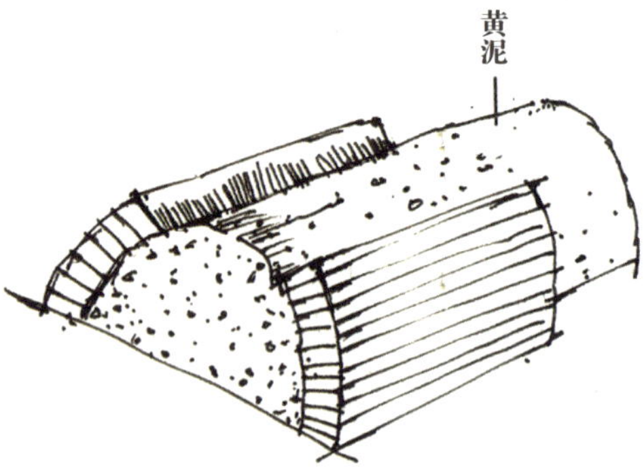

土砌法

木砌法

当地拱砌筑方法

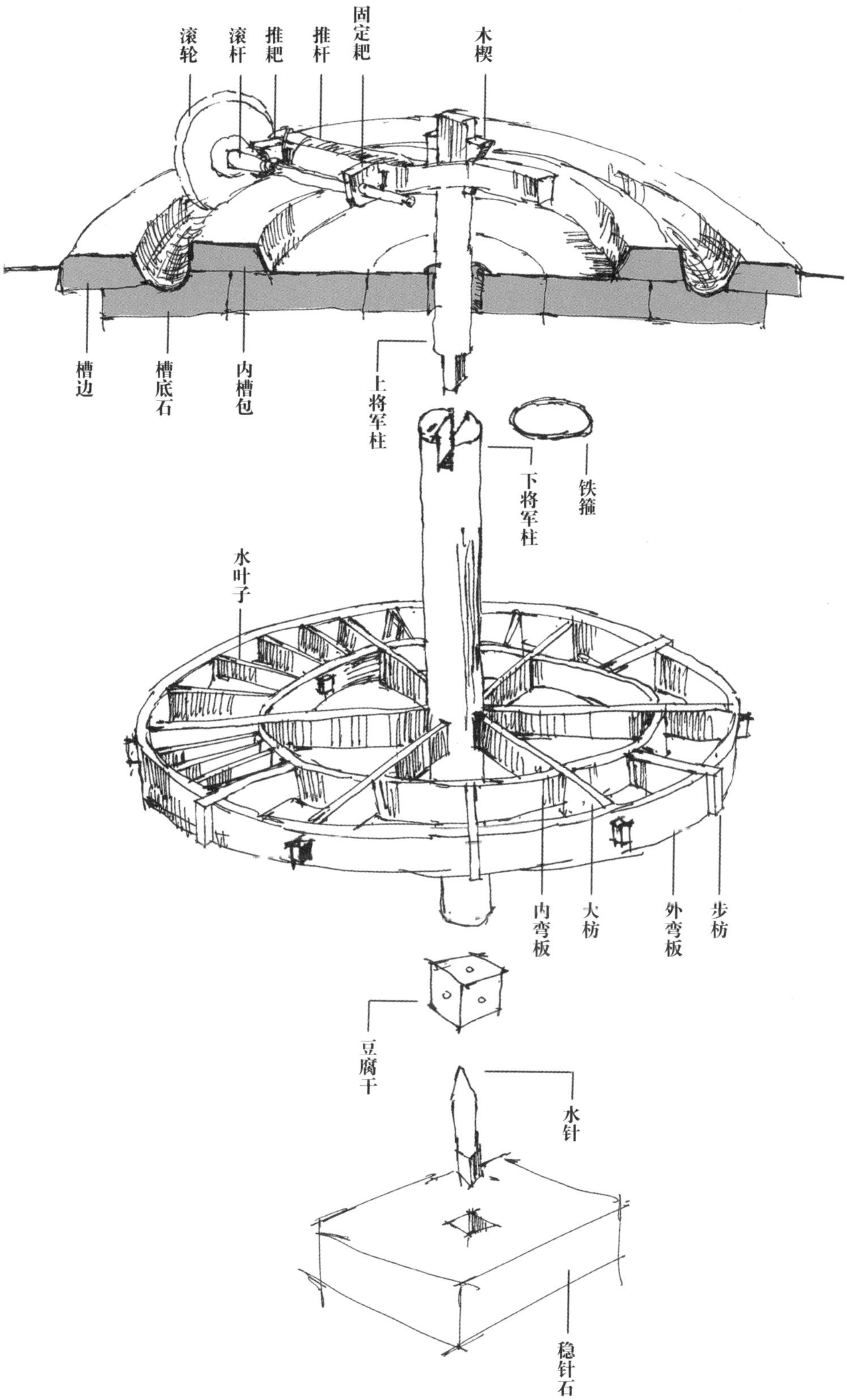
滚轮
滚杆
推耙
推杆
固定耙
木楔
槽边
槽底石
内槽包
上将军柱
下将军柱
铁箍
水叶子
内弯板
大枋
外弯板
步枋
豆腐干
水针
稳针石

水碾与水伞构造详图

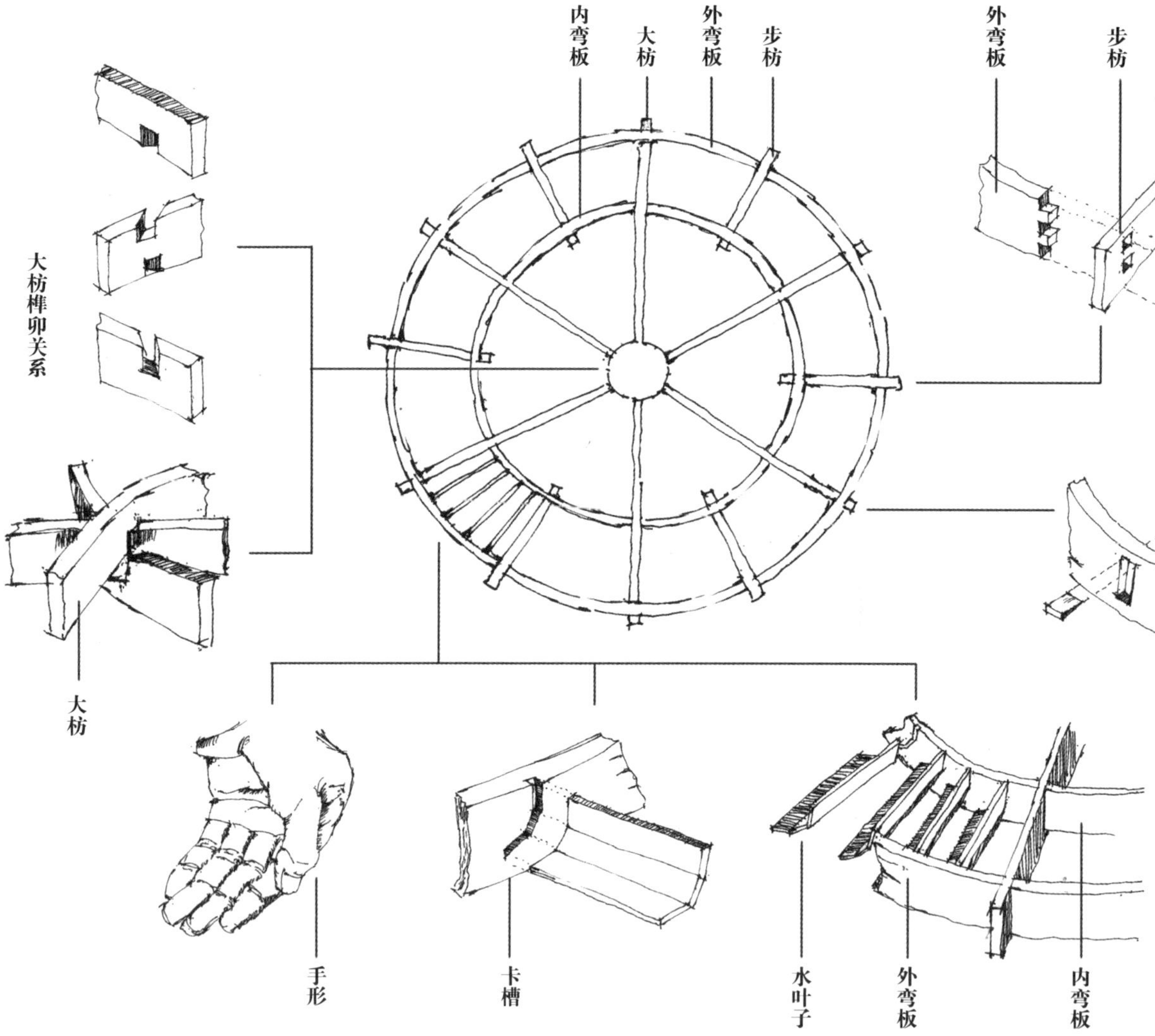

水伞构造大样

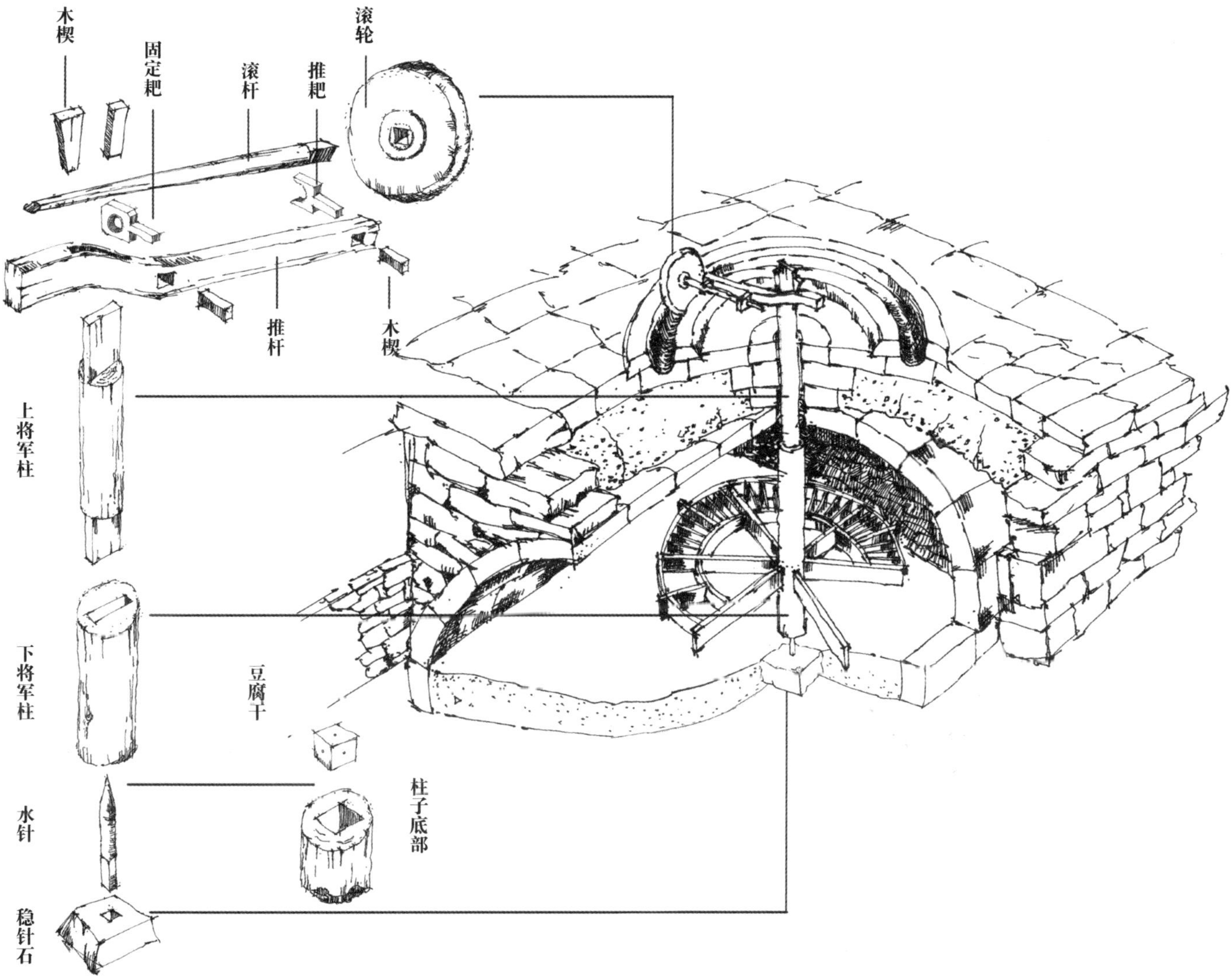

水碾与水伞透视图

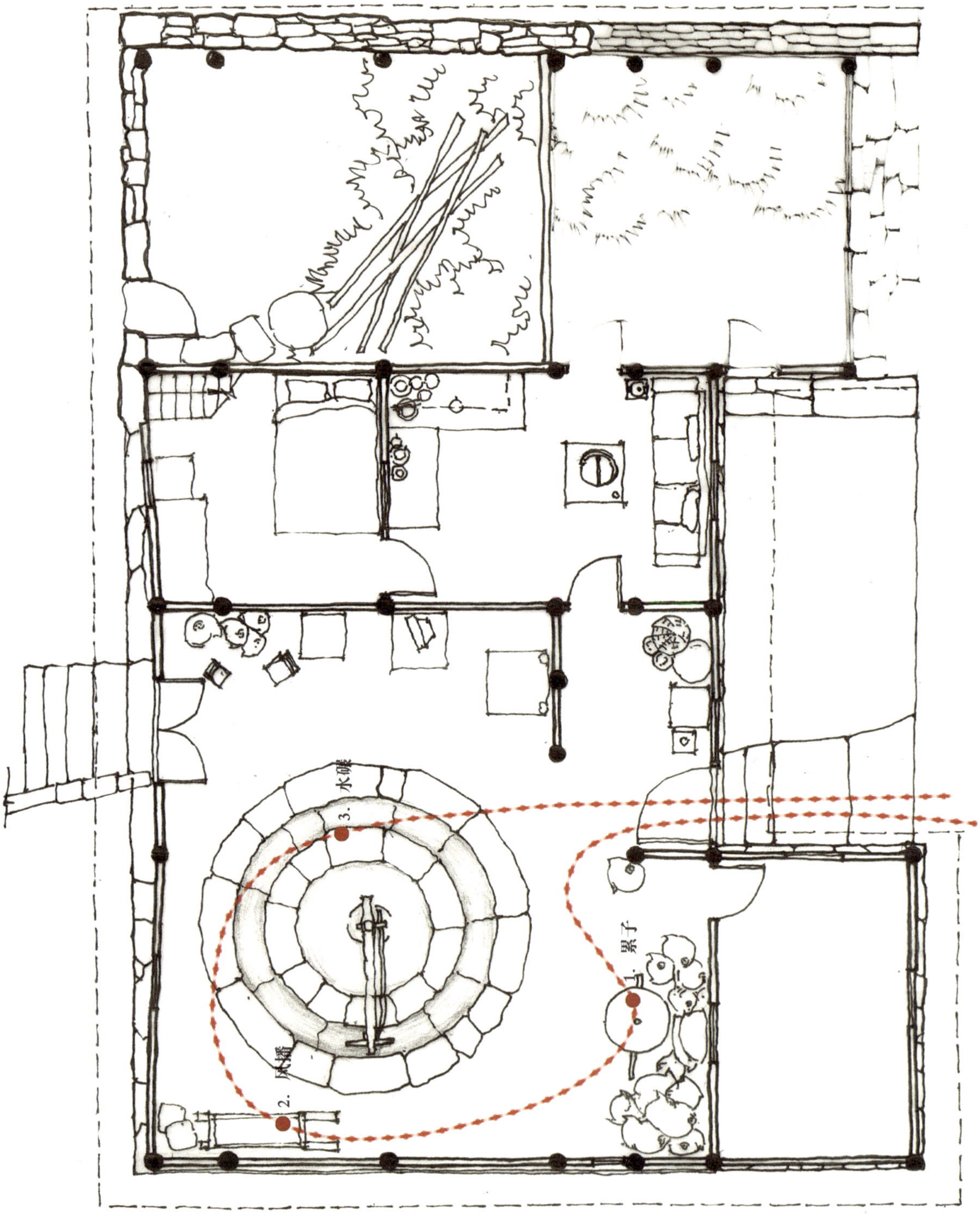

水碾使用流线示意

水碾日常使用图

水碾使用流线示意

7

8

9

三、鲍家屯水碾房修复工程申报书

an entry for

2011 UNESCO ASIA-PACIFIC HERITAGE AWARDS
for Culture Heritage Conservation

RESTORATION PROJECT OF BAOJIATUN WATERMILL

ANSHUN, GUIZHOU, CHINA

UNESCO
Asia-Pacific
Heritage
AWARDS
for
Culture Heritage Conservation

UNESCO
Culture in Asia and the Pacific

an entry for

2011 UNESCO ASIA-PACIFIC HERITAGE AWARD
for Culture Heritage Conservation

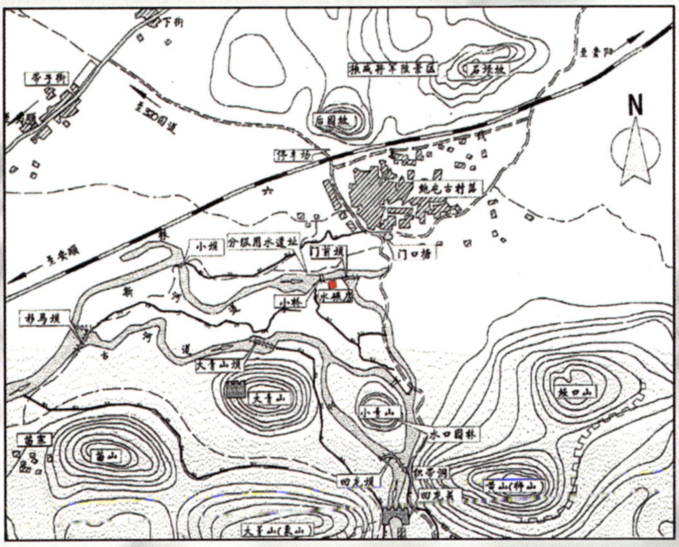

Restoration Project of Baojiatun Watermill

Anshun, Guizhou, China

UNESCO ASIA-PACIFIC HERITAGE AWARDS
for Culture Heritage Conservation

2011 ENTRY FORM

To: Culture Unit
United Nations Educational Scientific and Cultural Organization
UNESCO Bangkok Office
920 Sukhumvit Road, Bangkok 10110, THAILAND
Tel: (66 2) 391 0577 ext 509
Fax: (66 2) 391 0866
E-mail: culture.bgk@unesco.org

Date of Submission: **25 April 2011**

PLEASE TYPE:

Name of Project *(Entry)*

Restoration of *Baojiatun* watermill and related water conservancy facilities

Is your site a listed monument? Yes □ No ■

If yes, what year and what category?

Year:__________ Category:__________

Project is submitted for consideration to: (Tick ONLY one box – projects are not eligible for dual submission to both categories. Projects which are jointly submitted will be disqualified.)

■ 2011 UNESCO Heritage Awards for Culture Heritage Conservation

Location of Project *(Street, City, Postcode, State, Country)*

Baojiatun village, Anshun, Guizhou, China

Date of Completion of the Project *(must be after October 2001)* October 2009

Name or Names of Individuals Responsible for the Project

Professor Shixing Bao

Name of Building Owner whose permission for Nomination has been obtained *(if not submitted by owner)*

Mr. Zhongren Bao

Address of Owner

Baojiatun village, Anshun, Guizhou, China

Name(s) of Architect/Designer/Consultant

Professor Shixing Bao, Professor Qingzhou Wu, Senior Engineer Xuming Tan

Address(es) of Architect/Designer/Consultant

Shixing Bao 100055, Room 1503, Unit 1, Building 3, courryard #6, Maliandao Street, Xicheng District, Beijing

Xuming Tan 100044, No. 20, Chegongzhuang West Street, Haidian District, Beijing

Name of General Contractor

Mr. Zhilin Chen

Address of General Contractor

Baojiatun village, Anshun, Guizhou, China

SUBMITTED BY:

Name Shixing Bao

Address Room 1503, Unit 1, Building 3, courryard #6, Maliandao Street, Xicheng District, Beijing

City Beijing Postal Code 100055 Country China

Phone Number (86-10-63342866) Fax Number (86-20-87114884)

Mobile Phone Number (0086-13641038822) Email baosx403@sina.com

How did you learn about the Asia-Pacific Heritage Awards?

From Internet and Magazines

CHECKLIST:

The following documents are enclosed:

a. ■ Drawing, including site plan, floor plans and sections. (New, old, and reconstructed parts should be clearly marked in colour on the drawings). Total no. of enclosed:_______________

b. ■ 5" x 7" Photographs used to support the project description, clearly labeled with location and identified on a plan. Total no. enclosed: [min. 20]_______________

c. ■ Project Description (in English), 6 to 8 (A4 sheets):

d. □ Project Completion Report *(if available)*

e. ■ Owner Consent

f. ■ Occupant Comments

g. ■ Additional documentation *(not required)*

h. ■ CD-ROM with electronic versions of documents and photos (if possible)

NORTH VIEW:BEFORE and AFTER

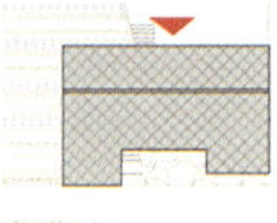

EAST VIEW:BEFORE and AFTER

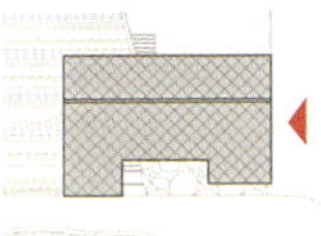

SOUTH VIEW:BEFORE and AFTER

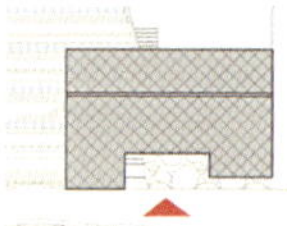

WEST VIEW:BEFORE and AFTER

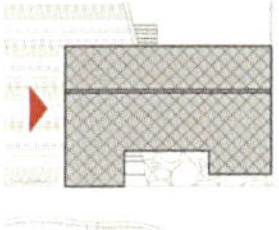

INTERIOR: BEFORE and AFTER

The covered roof by added wooden floor

The redone roof and the slide stone cover

INTERIOR: BEFORE and AFTER

The wasted stone mill

The reused stone mill

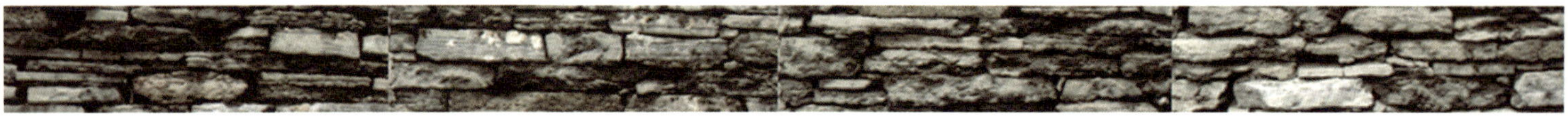

Restoration Project of Baojiatun Watermill, **Anshun, Guizhou, China**

• Synopsis of conservation project

In the north-east of An'shun town, Guizhou province, there is an ancient village. It is called "Baojiatun" because more than 90% residents share one family name "Bao". Baojiatun has a history of more than 600 years since the Ming Dynasty (1369). It originally existed as a garrison station of one vanguard of the Ming Dynasty Army in a historic war. When the war came to an end, the troop there did not leave but settled down, started farming and lived as regular farmers.

When Baojiatun was gradually being formed in the old ages, the ancestors built a dam to store floodwater for irrigating crops, and to overflow for flood discharge as well. By applying "fish-mouth dividing dike", a new canal (1.33 kilometers) which separated the upper river course was generated heading the lower reaches. Along the new canal, 5 diversion dams and 5 approach channels were built to irrigate the cropland. Later, 6 mills driven by water energy for grinding grain were built next to the 6 dams.

The object of the project is the only mill which is still in use now. It is the 1 of 6 which is the nearest to the village, with the longest history and best quality, linked the man-made canal and the entrance of the village. During the decades of historical vicissitudes, though the mill was fortunate enough to survive wars, it was still destroyed to different extents in many places.

The damages are mainly the structure parts, including the roof (broken), the west wall (collapsed), the windows and doors. It creates unsafe condition for the ordinary use of the mill. Gratifyingly, by the efforts of the experts together with the villagers, the mill was successfully restored in the year 2009. In the project, keeping and returning to its original appearance and function was placed as a high priority. For that, traditional techniques and local materials were used. Unsafe impacts were also eliminated by the restoration.

From 2009 to 2010, the mill and the surrounding water facilities facing the extreme droughty weather and successfully offer the village an efficient irrigation and production system. Furthermore, the potential ecological tourism may increase the village income to well protect the legacy and the traditional culture.

Left: The entrance of the Baojiatun village, the mill is on the left side
Right: The surrounding water facilities which link to the mill

Location of project in China

Location of project in Guizhou

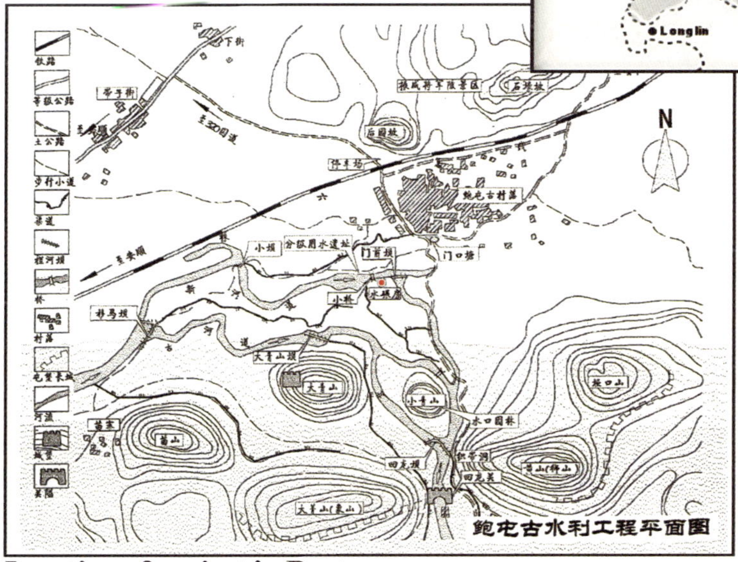

Location of project in Baotun

• Conservation project's goals and objectives (client's brief)

1.To display the history of Baojiatun and to establish the complete landscape system by restoring the watermill. The watermill is a main building in one of the 24 sight spots in the village, and it's on the left side of the only way for entering the village. Therfore, the restoration of the watermill is of great importance to the view and the history of the village.

As it showws in the pictures, the villagers in Baojiatun sill keep the tradition of playting JUN NUO for the worship of the God of water, to pary for the harvest. The social culture and the man-made water facilities is one of the most important parts to show the local activities and spirits.

2. To show the integrity of the water conservancy facilities in Baojiatun. The watermill is an integral part of Baojiatun's water conservancy system. There once were 6 watermills in history, but most of them were abandoned. In 3 of the remaining watermills, it is the only one that was well preserved and still functionable.

3. To play a positive and exemplary role in the conservation of nearby villages . Nowadays, some historic villages are so enthusiastic about modernization that they pulled down decorated archways and widen the main streets to let more vehicles in, and even changed the folk houses into big pseudo-classic style buildings, which completely changed the historical villages' original character . In comparison, this restoration project sticked to the principal of " rehabbing historical buildings as what they had been", which sets a silent example for the conservation work of neighbouring villages.

4. To ensure the sustainable use of the watermill. The watermill needs no fuel to drive, which conforms to current trend of energy-saving and low-carbon. As a result, the watermill should keep it's original function, and provide possibility of eco-tourism.

JUN NUO: a typical vernacular exorcise origins from the Ming Dynasty military event/festival. Still using for the worship of Beidi (God of Water) for the harvest

• The background of the property

• Historical background

In the year 1368, Ming Dynasty, ZHU Yuanzhang the emperor sent his troops to invade Guizhou province. The troops occupied everywhere they conquered for supply and logistics needs.

Bao Fubao, the leader of one troop, chose where Baojiatun current locates as his garrison station. After the war, they started farming, multiplied and lived there; and that is the origin of Baojiatun. Thus, their architectures and costumes reflect obvious traditional style of Ming Dynasty.

As soon as Bao Fubao chose the place for his troop, he brought rich knowledge and experience of water conservancy from his hometown (South of Anhui). A systemic water conservancy work was established gradually. By fully considering the local condition, the work was designed to separate the river into two, and formed several circuitous sub-canals, which could irrigate all cropland (30 km2) in different elevation. Plus, there are 6 mills driven by water energy next to the dams.

But owning to the impacts of the various social changes and the misuses, most mills were seriously damaged while only one survived and is still capable of working.

The damage condition before the restoration was terribly bad.

1. **The roof** of the building was broken, some parts were substituted by tiles but nor original slice stones, some parts were covered by asbestos. It was difficult for the users to keep it as rain-proof cover, and also difficult to keep the room warm in the cold weather.

2. **The walls** of the mill was held by the wooden structure. Most the wooden parts were inclined. It created the lean of the east wall and the collapse of the west wall. Also it made the roof of the building in a curve shape which may result the slide or the fallen of the slice stones.

3. **The windows and the doors** were damaged. They couldn't fit the structure of the flame and hardly been closed.

4. **The short wall** outside the mill to separate the milling wheel and the outside canal were partly collapsed. It suffered the unsafe situation during the rain and flood season. Also the fallen of the stones and the gaps between the stones in the short wall might be a crucial reason for the terrible humility inside the mill.

In oder to preserve the last watermill as a witness of Baojiatun's past glory, the restoration project was initiated in 2008.

With many efforts, the mill and its relevant facilities were successfully restored and are now used in good conditions .

• Present context, condition, use of the property

The mills are important components of the water conservancy work in Baojiatun. It located in the middle of paddy cropland; with river coming from the north, courtyard in the south, and stone bridge in the north-west. The orient part of the mill is above the river. When water in the river fell down from north to south through the mill, the waterwheel below was put in motion to drive the roller for grinding grain.

By the restoration, the mill now works well with the grain roller. And it is even available for villagers living there.

• Heritage/protection status of the property

Being worn down by the years without repair, the mill was in poor condition that it could barely work:

• *Appearance*

The central room: luckily, all the doors,steps and the grain roller inside the room were preserved in fair condition.

The room in the west wing: the west facade was kept relatively intact.

The room in the east wing: the envelops were constructed with coal-dust bricks but lost protection from wooden boards, which let its chuandou system(a special kind of structure in traditional Chinese architecture, which is also called "through-type timber frame" or "column-and-tie construction". In this system, people use tall, thin roof purlin-to-ground columns along the full length of the gable end and horizontal tie beams that penetrate these timber columns.) exposed.

The roof: a few worn-out slates in the east.

PS: In each facade, it is all too easy to find rotted wooden boards or damaged stone slabs with joints filled with various modern materials .

• *Structure*

The chuandou system inclined towards west, which put the whole building in danger.

• *Water conservancy facilities*

The waterwheel and grain roller are kept in good condition and still in use.

These problems have already been settled after restoration.

The property of the mill belongs to the villager of Baojiatun. The Village commitee takes care of the normal usage and maintance after the restoration.

In 2010, Baojiatun, the whole village, is nominated as the National Historic Village, and the water facilities and the mill is applying for the provincial level and state level Culture Relic, in the future, it may also under jurisdiction of relevant culture and construction department, and enjoys periodical maintenance.

Besides, the water system of Anshun city is also under the management of the prefectural hydraulic department, regular inspection and maintance of the water facilities in Baojiatun village are also taken care of by them.

• Scope of the conservation project

• Stages in restoration and time frame

Working Progress	Time
Reaserch	2008.2.~2010.3.
Mapping	2008.3.~2010.5.
Restoration	2008.9.~2008.12.
Final Acceptance	2009.10.

• Project size (in square meters) and cost (in US$)

Project	Size
Dike Restoration	2623m
Reshaping of Irrigation Channels	7 channels,with total length of 4080m
Reshaping of Channel branches in field	3500m
Dam Restoration	3
Watermill Restoration	1

Funding	Amount
Research fund from China Institute of Water Resources and Hydropower	USD4,587

• Articulation of heritage values and significance

The mill is the key part of the whole system in Baojiatun water facilities.

From a historic point of view, the mill keeps the original part of the water facilities built in the history, which is still functionable. The restoration makes the culture of using nature resources to be the energy of the daily life. The fallen water caused by the dam drives the roller in the mill for grilling grain. The traditional sustainable philosophy is simple but very efficient.

From an architectural point of view, the mill keeps the vernacular tectonic methods and materials. The local people can still work properly to do the restoration project. Also the professional designers help to instruct and record the progress of the project.

From a social point of view, after 600 year's mutual adaptation, the water conservancy facilities have been well harmonized with the local environment and ecology. It helps the canal system for the second-phase of production.

From a disaster proof point of view, the mill connects the canals, and creates a whole system of dams, dykes and water gates. During the terrible draughty and flood disaster, there are no severe negative influences in Baojiatun thanks the water conservancy work. For the daily life, the capable coverage of the irrigation is 153.3 ha. It is very efficient by using the dams and canals. Yielding two crops a year is possible, like paddy, wheat, rape and broad bean, etc.

• Appropriate use/adaptation

• Changes made to building (repairs and additions), and relationship of new to old

• *Repairs*

1.Repair of main structure(chuandou system)

The chuandou system inclined towards west before restoration. In order to set it right, we detached components of the structure one after another, and then assemble them back in order.

2.Repair of building envelops(including walls and the roof)

Walls:The methods of adding enclosure walls to the main structure (chuandou system) varies at different heights.In the plinth, we used stone strips to prevent flood; under window-sills, we adopted local material horizontally-sliced stone slabs(stone slabs which are from a kind of gneiss that are common in Baojiatun. Through ages, the gneiss was natrually sliced into horizontal slabs, some of which--ones with good quality and certain thickness--are picked out and set upright to serve as interior partitions or enclosure walls under window sills)to enclose; between the windows, wooden boards are butted together to serve as external walls; and above the windows, we adopted a special form of wall that is called zhuba-wall, in which bamboo splints are tightly weaved together to fit in the wall's basic framework and lime mortar mixed with cow dung is smeared on surface as plastering.Besides, as the river flows from west, we added a stone wall whose upper edge is even with the eaves on west facade in case of flood.

The roof:slates with naturally irregular shapes overlap each other--like the way fish scales fit together-- to form the roof.

3.Repair of water conservancy facilities:As the water wheeler and the grain roller were kept almost intact, we just left them as what they were.

4.Repair of decorations: Restore defective traceries refer to the well-preserved one of west room and without any new addition.

• *Relationship of new to old:* The project made full use of old materials in the building. Even when new materials were required in some damaged parts, they were reworked into antiquated forms, which brought a harmony relationship of new to old.

1. Except the wooden structure are partly replaced because of the rotted condition, the others are kept as much as possible the original statues. The lean columns were adjusted to hold the renewed roof and wall.

2. The west wall are reconstructed with local materials and the same tectonic methods of the left east wall, covered by bamboo splints in the mid and cow dung in the face (for water-proof).

3. The roof is redone after the adjustment of the columns. The project uses the same stones, which are both from local mountains, cut into slides, to cover the roof. It is also a traditional and vernacular tectonic of building constructions.

4. The windows and doors are renewed with the same decoration of the left ones. The wooden material is also same as the original parts.

• Interpretation of the significance of the building(s) in the conservation work

• The methodology of restoration

1. Safety first.
2. Restore to what it was.
3. Apply traditional technique.
4. Use original and local material for construction as far as possible.
5. Consider sustainability.

• Understanding of technical issues of conservation in the interpretation

• Issues of conservation/restoration considered and justifications for solutions

1.Overall architectural layout

Under the guidance of the principle that restore historical buildings as what they had been,we developed following instructions:

Keep down the building's original chuandou system but with unsafe factors removed.

Continue the property's function as a watermill.

Use local materials; give priority to existing materials in the building, and if any new materials are added, they should be reworked into antiquated forms.

Adopt traditional techniques which will not artificially change irregular shapes of materials.

Restore all the decorations in accordance with their original patterns.

2.Needs for Structural Safety and Normal Uses

The main structure(chuandou system) needs to be righted , for it had already inclined ,which put the whole building in danger.

Add eaves-even-wall (a stone wall whose upper edge is even with the eaves. it is placed against the external wall to reinforce the house and to prevent flood) on west facade (in the direction opposite to the stream) .

Continue the building's function as a wtermill and at the same time provide housing for villagers.

3.Material Selection and Sourcing

Use local materials and try to avoid modern building materials such as cement.

Take advantage of used materials if possible(but be aware that some rotted or fractured timbers must be replaced for safety).

In terms of material processing, for timbers, clean dust out first, then polish with sandpaper, and at last brush over with anticorrosive tung-oil; And for masonry, all the stones need to be reworked into antiquated forms and bound together with lime mortar(dry-brick building can also be used in non-load bearing parts).

4.Ornaments

Restore defective traceries refer to the well preserved ones in the west wing.

• Use of appropriate building, artisan, and conservation techniques

All the techniques adopted in this project are strictly in accordance with Baojiatun's traditional crafts.
Specific techniques are as following:

• *Horizontally-sliced stone slabs:* As Baojiatun is abound in a kind of gneiss that was natrually sliced into horizontal slabs through ages, we picked out some slabs with good quality and certain thickness, and set them upright to serve as interior partitions and enclosure walls under window sills.

• *Eaves-even-wall:* Learning from local conventional practice , we placed a stone wall whose upper edge is even with the eaves against the external wall to reinforce the house and to prevent flood.

• *Roof slates:* Slates with naturally irregular shapes overlap each other--like the way fish scales fit together-- to form the roof.

• *Zhuba-wall:* We adopted a special form of wall that is widely used in local housing.To form this wall, bamboo splints are tightly weaved together to fit in its timber framework and lime mortar mixed with cow dung is smeared on surface as plastering.

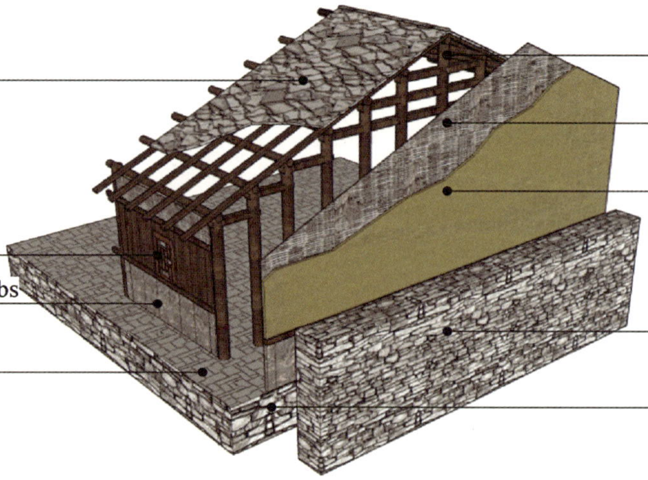

The technical names of local materials

Chuandou system

A special kind of structure in traditional Chinese architecture, which is also called "through-type timber frame" or "column-and-tie construction".

In this system, people use tall,thin roof purlin-to-ground columns along the full length of the gable end and horizontal tie beams that penetrate these timber columns.

Horizontally-sliced stone slabs

Stone slabs which are from a kind of gneiss that is common in Baojiatun. Through ages, the gneiss was natrually sliced into horizontal slabs, some of which--ones with good quality and certain thickness--are picked out and set upright to serve as interior partitions or enclosure walls under window sills.

Zhuba-wall

A special form of wall in which bamboo splints are tightly weaved together to form the wall's basic framework and cow dung is smeared on surface as plastering.

Roof slates

Slates with naturally irregular shapes that overlap each other--like the way fish scales fit together-- to form the roof.

Eaves-even-wall

A stone wall whose upper edge is even with the eaves. it is placed against the exterior wall to reinforce the house and to prevent flood.

The restoration progress of the mill
It shows the Chuandou system clearly

• Use of appropriate materials

• Timber:Used in structure and some partitions. Specific techniques are as follows: clean dust out first, then polish with sandpaper, and at last brush over with anticorrosive tung-oil.

• Slates:Used for roof. Keep slates' irregular shapes and overlap them as the way fish scales fit together.

• Stones:Used in walls and the plinth. New stones need to be reworked into antiquated forms and bound together with lime mortar. Besides, dry-brick building can also be used in non-load bearing parts.

• Bamboo splints & cow dung:Used to constitute zhuba-wall which is placed in the end wall. To form this wall, bamboo splints are tightly weaved together to fit in its timber framework and lime mortar mixed with cow dung is smeared on surface as plastering.

Chuandou system

Lattice window

Roof slates

Eaves-even-wall

Horizontally-sliced stone slabs

Bamboo splints

Cow dung

• How well new elements and creative technical solutions respect building's character

To show respect for the building's character,we adopted only local materials and constructed strictly in accordance with traditional techniques. Any new elements or modern technical solutions are excluded.

• Contribution to community's cultural continuum

• Key stakeholders and involvement of local community in project

1. *Researchers and designers involved:* Professor Shixing Bao, Professor Qingzhou Wu, Senior Engineer Xuming Tan.

2. *Director of construction:* Baojiatun villagers committees . Bao Zhongren, the village chief, took charge of the process.

3. *Lead craftsman:* Chen Zhilin, the villager from Baojiatun village, took charge of the repairment of wooden structure, roof surface, walls, wooden windows, doors and interior structures.

• Impact of project on conservation practice and policy

Baojiatun has successfully applied for the province-listed historic and cultural village.
The apply of the state-listed historic and cultural village was approved in 2010.
The ancient water conservancy in Baojiatun has applied for the 7th batch of important cultural relic sites under state-level protection and received high opinions from scholars.

The restoration method of keeping them as what they are plays a positive and exemplary role in the conservation of nearby villages.

The hydraulic elements of the mill:
Water gates, Water mouth (for flood control), water-power mill wheel

• Ongoing socio-economic viability and relevance of the project

• Regulatory framework of the project

1. International Documentation on Historical Heritage Conservation such as Nara Document on Authenticity, Venice Charter and Charter on the Built Vernacular Heritage

2. Law of the People's Republic of China on Protection of Cultural Relics

3. The Law of the P. R. China on Land Management

4. The Law of the P. R. China on Rural Land Contract

5. Regulations for the Implementation of Urban Planning Compiling Method

6. Town and Village Planning Compiling Method

7. Urban State-owned Land Use Right Transfer Management

8. Township Planning and Construction Management

9. Standard of Village and Town Planning (GB50188-93)

10. Regulations of Rural Homestead Land Property Register

11. Regulations for the Implementation of the Law of the P. R. China on Urban and Rural Planning in Guizhou

12. Regulations of Rural Residential Planning in Guizhou

13. Regulations of Urban Institutional Reform in Guizhou

14. Regulations of Village Planning in Guizhou

15. Regulations of Rural Collective Land Use Right Transfer in Guizhou

16. Regulations of Rural Homestead Land Management in Guizhou

17. Regulations of Village Planning in Guizhou

• Mobilization of financial and in-kind support from public and private sectors

Funding: USD4,587 from China Institute of Water Resources and Hydropower*
Non-cash support:
Gratis construction management from Baojiatun villagers committees
Gratis technical support from local craftsman

** China Institute of Water Resources and Hydropower is a government department that manage the water and hydrulic works. The fund is a research support fund from this insitite, to encourage the scholars from academic insititute to do researches on the related topics. The money (RMB30,000 yuan, around USD4,587 dollars) is the part of the research fund that supprts the study of the water system in Anshun area in the local universities.*

• If project is a public-private initiative, contributions of each partner should be clearly indicated

Contributions of each partner:
China Institute of Water Resources and Hydropower:Provide funding amounting to USD4,587 in total.
Researchers and designers involved:Professor Shixing Bao, Professor Qingzhou Wu, Senior Engineer Xuming Tan. They were in charge of designing and instructed construction.
Baojiatun villagers committees:Be responsible for construction

• Provision for future use and maintenance

Provision for future use:
The Watermill shall be used for grain milling and as residence for local villagers in the meantime.
Provision for maintenance:
Under the jurisdiction of the department in charge of cultural relics management*;
Regular examination(about once a month) by local craftsman.

** Baojitun Village is on the list of the National level Historic Villages, which is managed by the construction departments of central government and local government. The mill and the water facilities is also applying for the National level Culture Relics. After the approvement, it will be also under the control, maintenance from the cultural departments.*

Websides and newspaper reports about Baojiatun and the water facilities

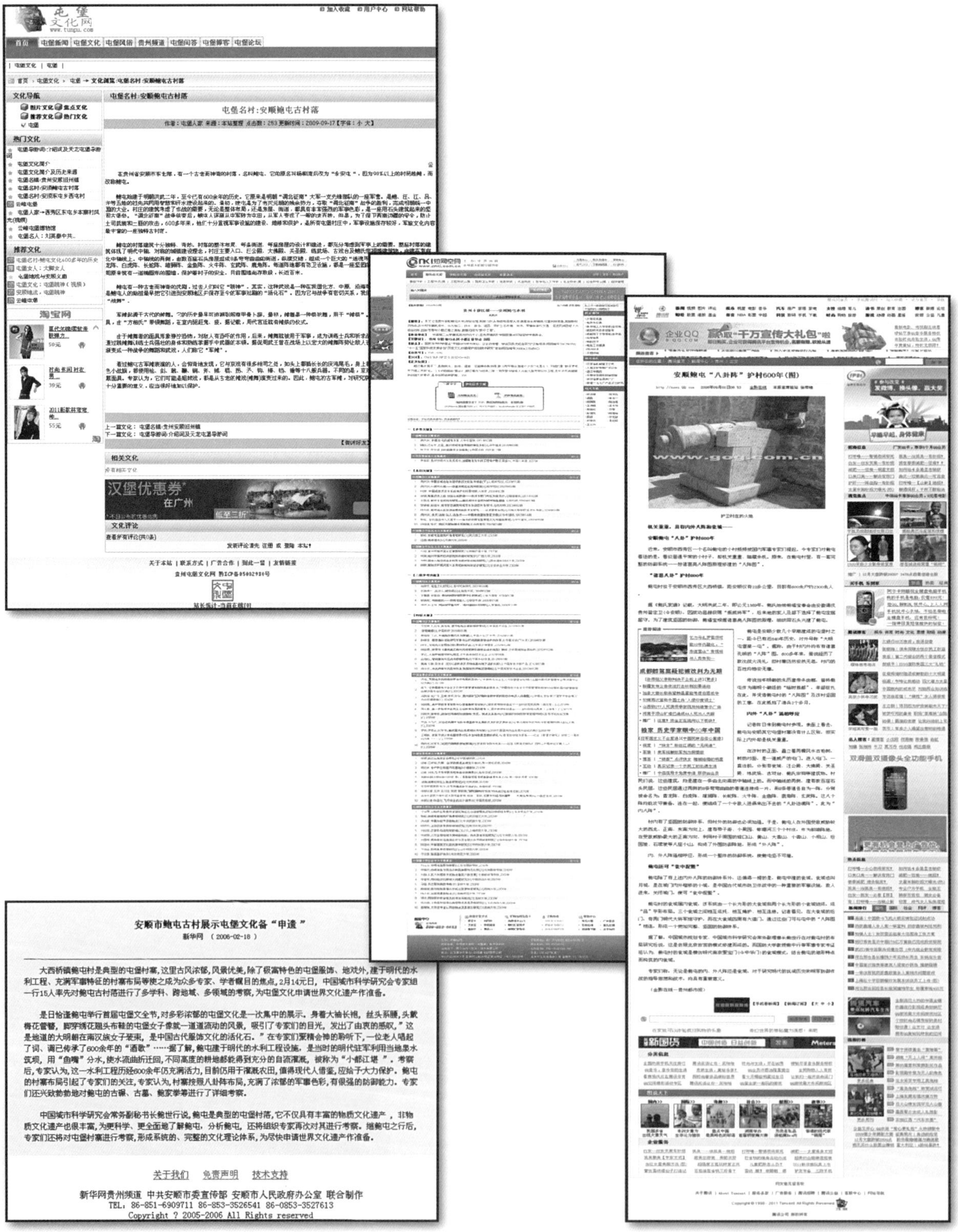

安顺市鲍屯古村展示屯堡文化备“申遗 ”

新华网 （ 2006-02-18 ）

大西桥镇鲍屯村是典型的屯堡村寨,这里古风浓郁,风景优美,除了极富特色的屯堡服饰、地戏外,建于明代的水利工程、充满军事特征的村寨布局等使之成为众多专家、学者瞩目的焦点,2月14元日，中国城市科学研究会专家组一行15人率先对鲍屯古村落进行了多学科、跨地域、多领域的考察,为屯堡文化申请世界文化遗产作准备。

是日恰逢鲍屯举行首届屯堡文全节,对多彩浓郁的屯堡文化是一次集中的展示。身着大袖长袍，丝头系腰,头戴梅花簪簪，脚穿绣花翘头布鞋的屯堡女子像就一道道流动的风景，吸引了专家们的目光，发出了由衷的感叹,”这是地道的大明朝在南汉族女子装束，是中国古代服饰文化的活化石。”在专家们聚精会神的聆听下,一位老人唱起了词、调已传承了600余年的“酒歌”……据了解,鲍屯建于明代的水利工程设施，是当时的明代驻军利用当地渠水筑坝，用“鱼嘴”分水,使水流曲折迂回,不同高度的耕地都能得到充分的自流灌溉，被称为“小都江堰 ”。考察后,专家认为,这一水利工程历经600余年仍充满活力,目前仍用于灌溉农田,值得现代人借鉴,应给予大力保护。鲍屯的村寨布局引起了专家们的关注,专家认为,村寨按照八卦阵布局,充满了浓郁的军事色彩,有很强的防御能力。专家们还兴致勃勃地对鲍屯的古碾、古墓、鲍家拳等进行了详细考察。

中国城市科学研究会常务副秘书长鲍世行说,鲍屯是典型的屯堡村落,它不仅具有丰富的物质文化遗产 ，非物质文化遗产也很丰富,为更科学、更全面地了解鲍屯，分析鲍屯，还将组织专家再次对其进行考察。继鲍屯之行后,专家们还将对屯堡村寨进行考察,形成系统的、完整的文化理论体系,为尽快申请世界文化遗产作准备。

关于我们 免责声明 技术支持

新华网贵州频道 中共安顺市委宣传部 安顺市人民政府办公室 联合制作
TEL：86-851-6909711 86-853-3526541 86-0853-3527613

• The complexity, sensitivity and technical consistency of the project methodology

The project aims to eliminate unsafe factors of the surviving watermill and furthermore develop sustainable use of its hydro-facilities in various ways such as eco-tourism. Thus more concern needs to be expressed for the complexity, sensitivity and technical consistency in the whole process .

• Complexity

Aside from protecting the historical building, the project also works at spreading traditional culture and creating certain economic values by tourism development, which brings about complexity when moving forward.

• Sensitivity

1.Sense of history in space

The water conservancy facilities remain valid and will provide possibility for touring while they. work as before. The respect for heritage's original use will certainly add a sense of history in spatial perception.

2.Sense of authenticity in decoration and facilities

Reduce visual impact of inevitable adding facilities(e.g.illumination and fire control facilities) ,leaving them only identifiability .

New decorations are forbidden ,except for some movable components for exhibit.

3.technical consistency

Sticking to the principle of rehabbing historical buildings as what they had been, we adopted local materials and traditional techniques. And in the working progress, we studied existing samples and consulted local craftsmen for technical details before construction.

Baojiatun Culture Festival, worship to the God of Water

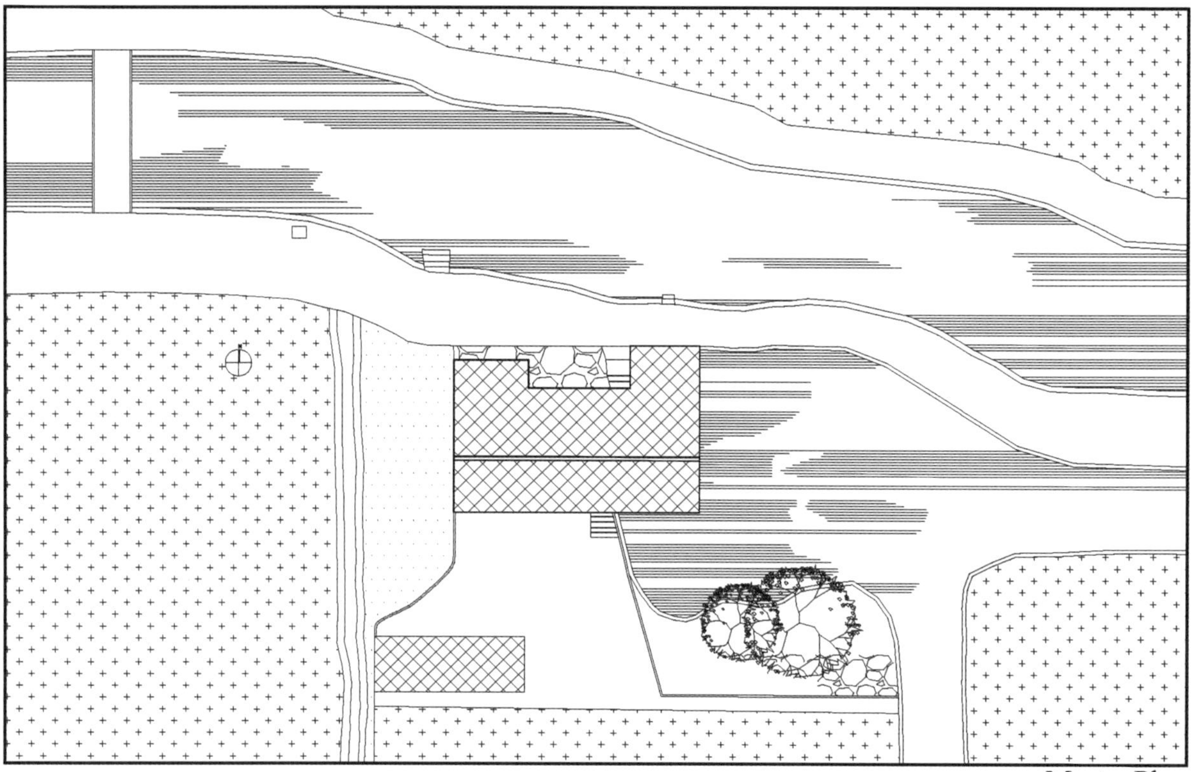

Master Plan

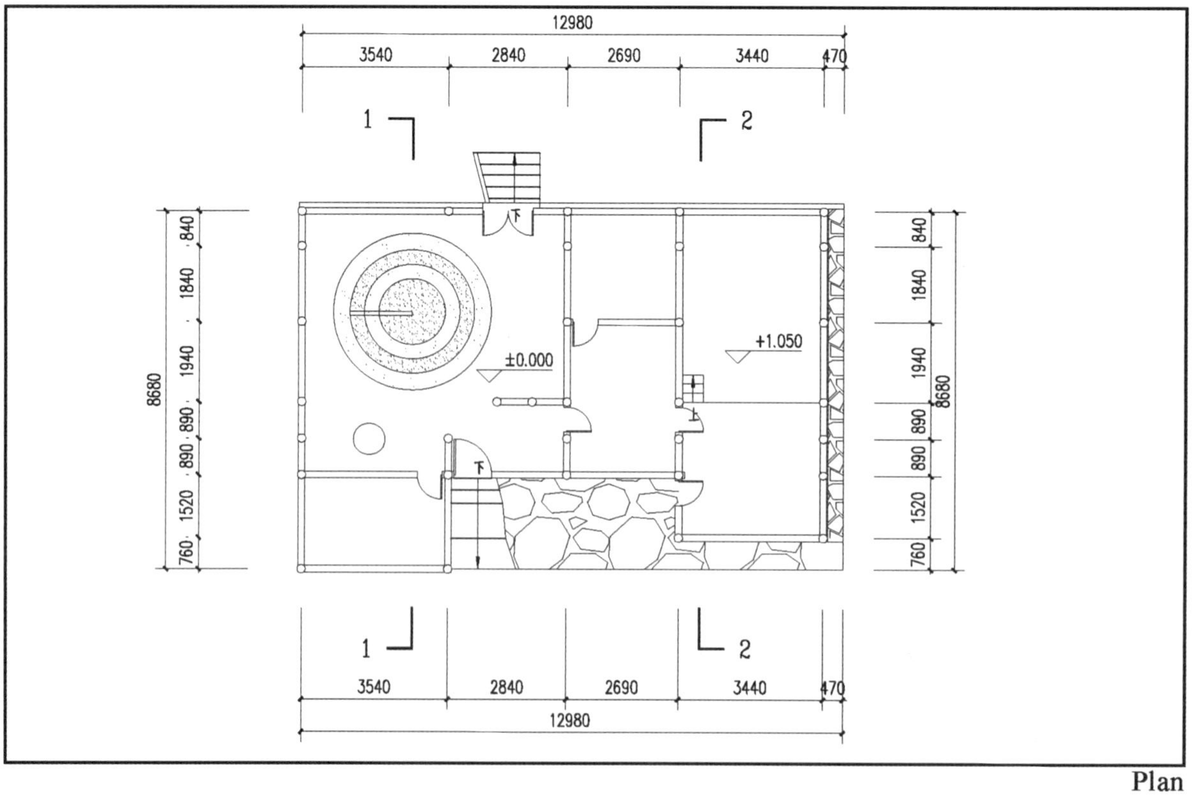

Plan

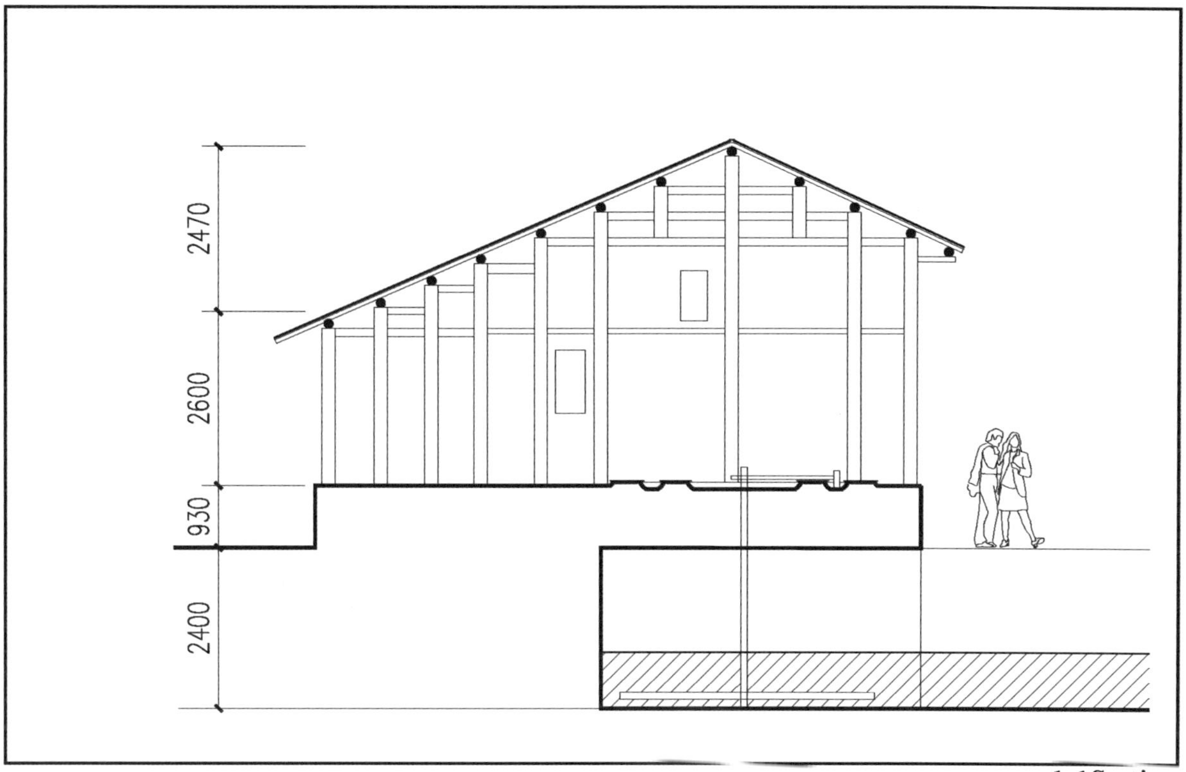

1-1Section

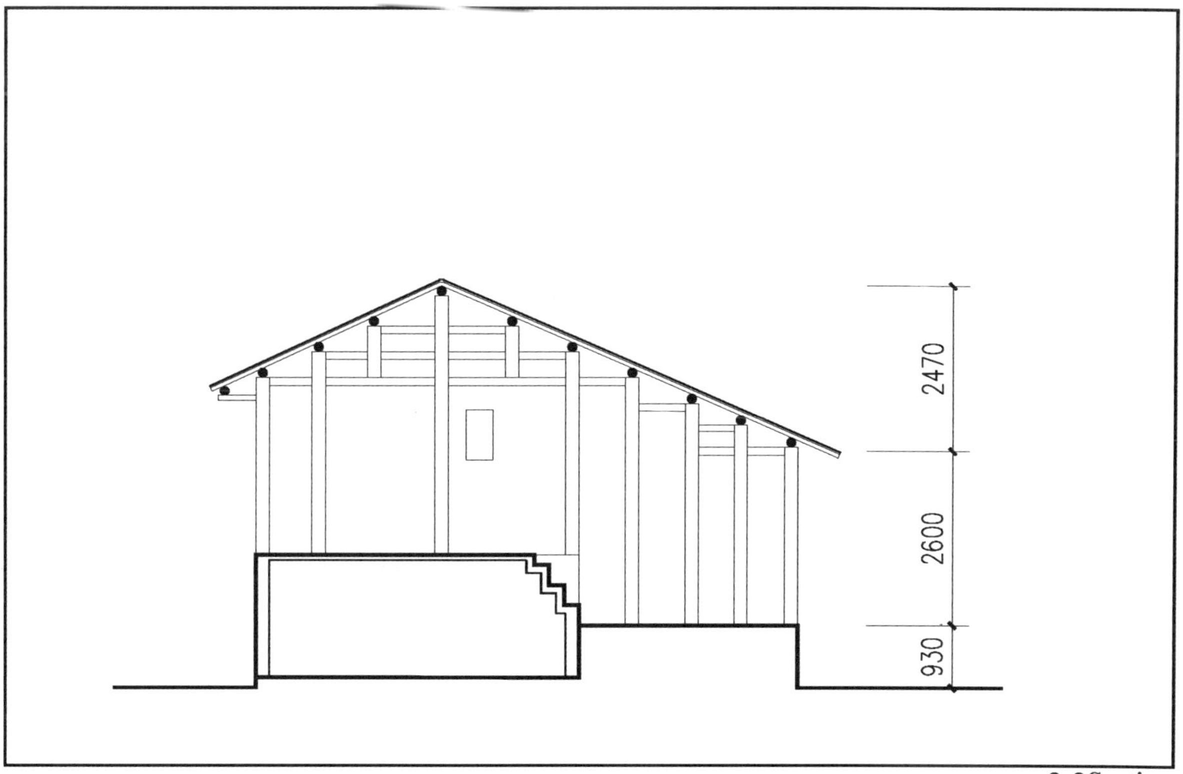

2-2Section

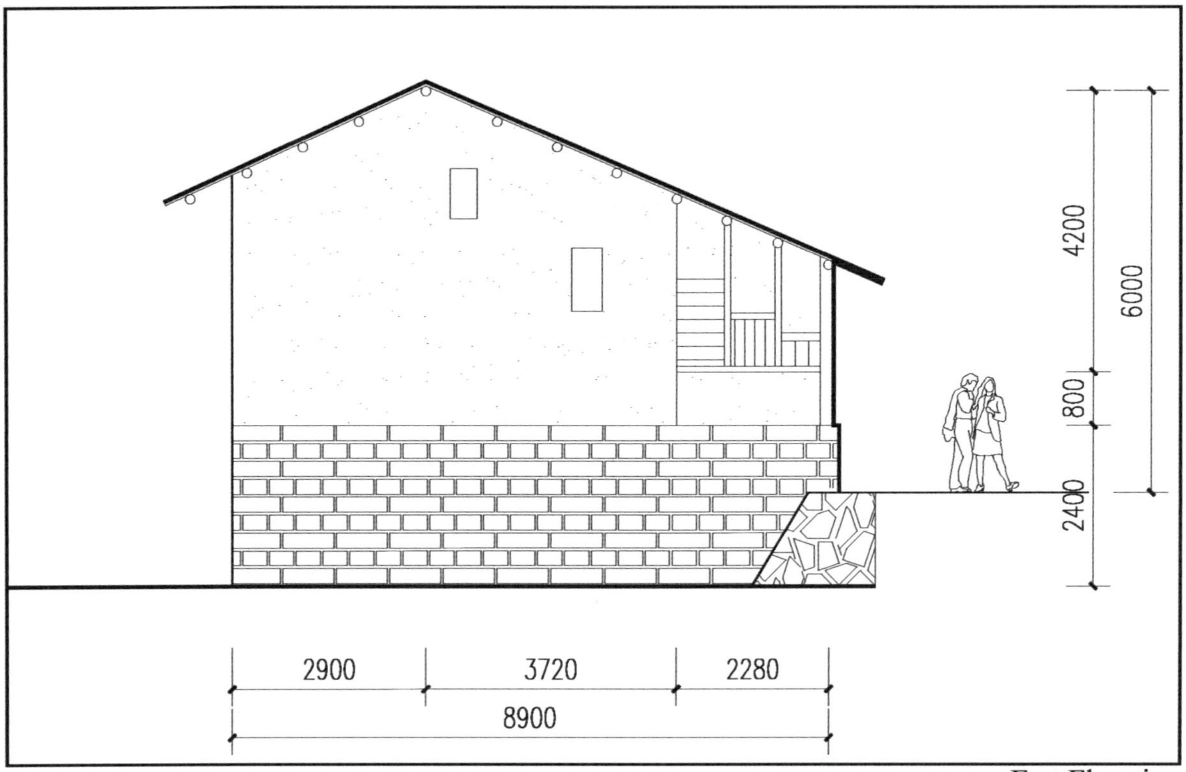

Eest Elevation

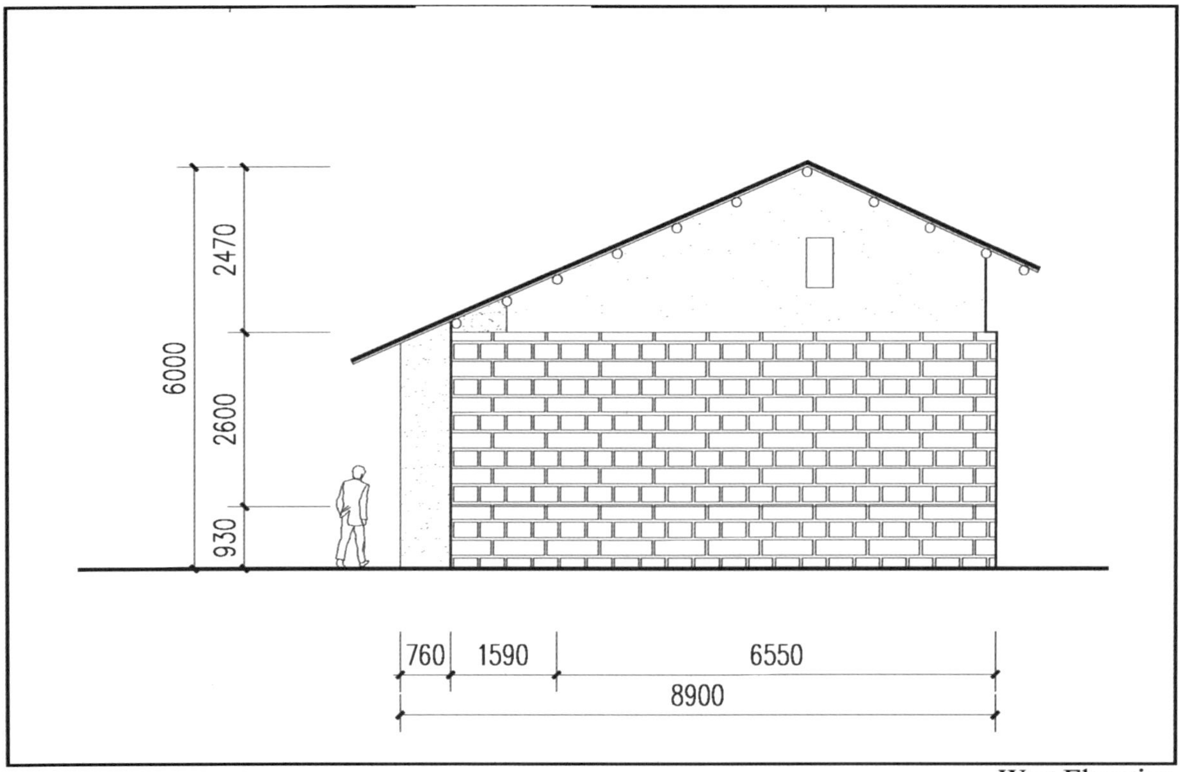

West Elevation

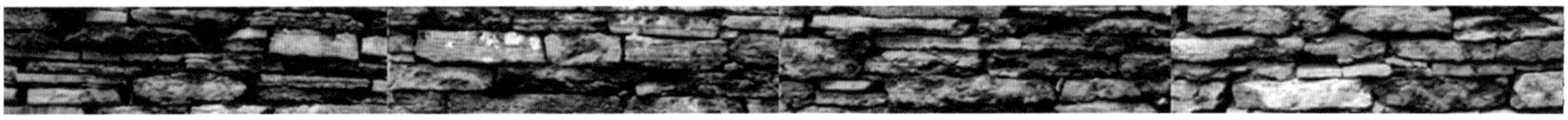

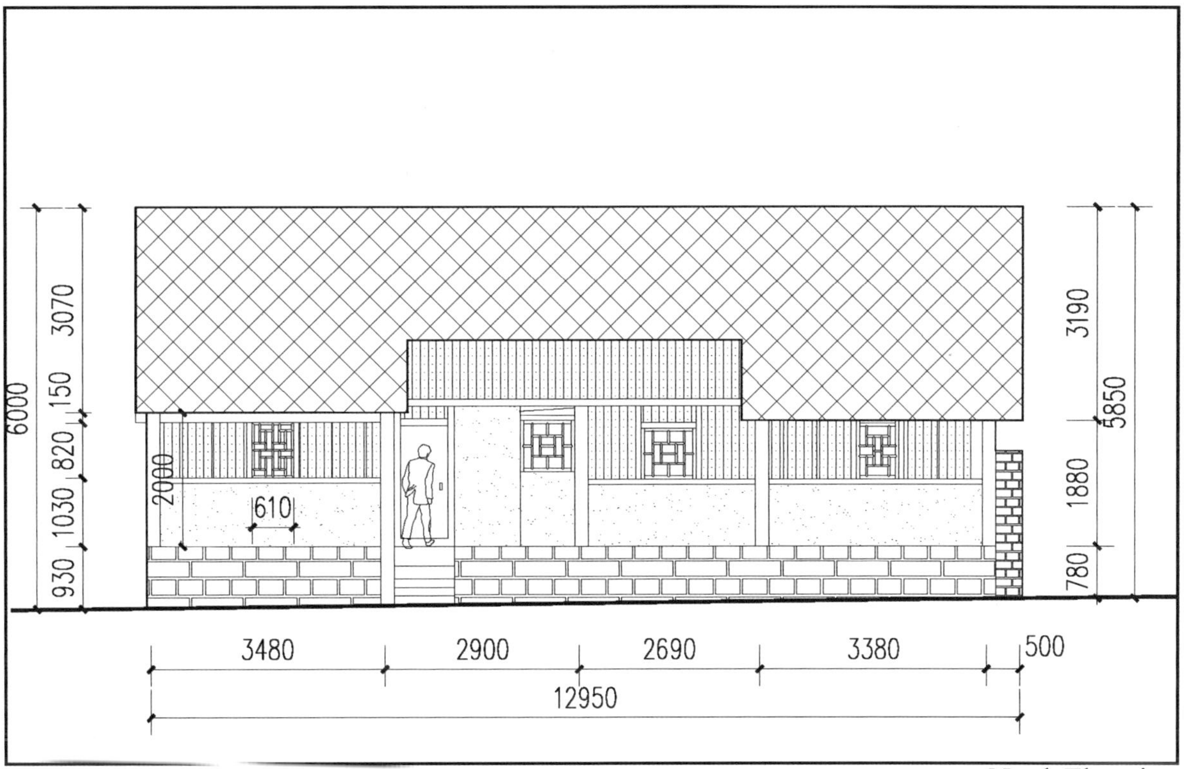

North Elevation

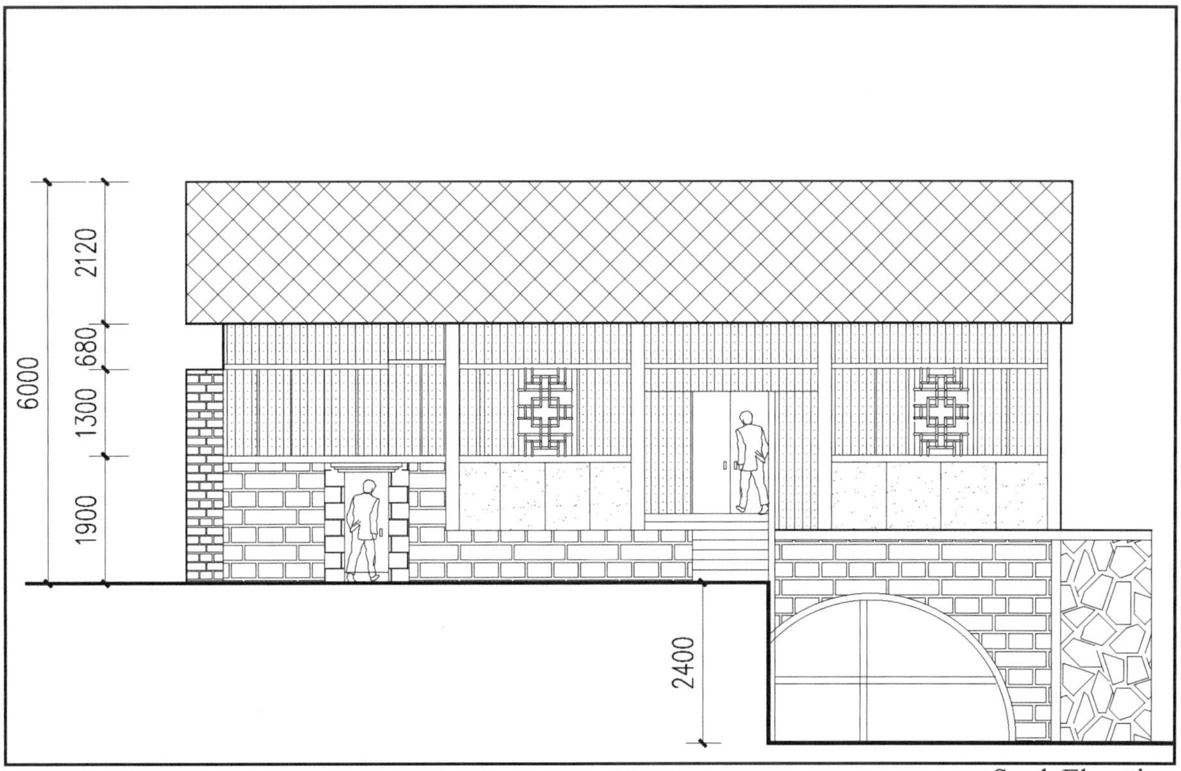

Sorth Elevation

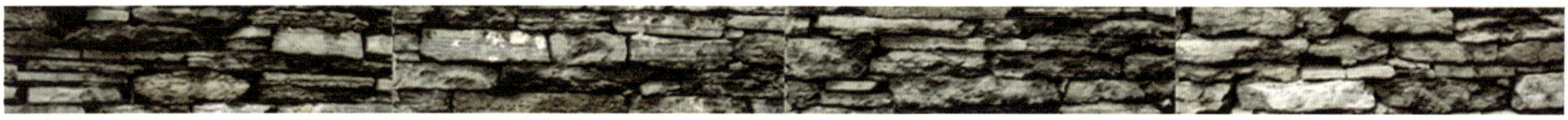

Restoration Project of Baojiatun Watermill, **Anshun, Guizhou, China**

UNESCO Asia-Pacific Heritage AWARDS for Culture Heritage Conservation

2011 UNESCO Heritage Awards for Culture Heritage Conservation

OWNER CONSENT FORM

20 April 2011

Culture Unit
UNESCO Bangkok Office
Bangkok, Thailand

Dear Sir:

The project of Restoration of Baojiatun Watermill and its Water Conservancy Facilities, (Anshun, Guizhou, China)

This is to certify that the above building is the property of Baojiatun Village. I hereby grant permission to Prof. Shixing Bao to nominate the above building for the 2011 UNESCO Asia-Pacific Heritage Awards for Culture Heritage Conservation.

安顺鲍家屯水碾房及水利设施修复项目
鲍家屯水碾房属鲍家屯村村民产权。我以鲍家屯村村长名义，据此同意鲍世行教授代表我们申请 2011 年联合国教科文组织亚太区文化遗产保护奖。

Sincerely yours,

鲍中任
Zhongren Bao

Baojiatun village, Anshun, Guizhou, China

UNESCO
Asia-Pacific
Heritage
AWARDS
for
Culture Heritage Conservation

2011 UNESCO Heritage Awards for Culture Heritage Conservation

OCCUPANT COMMENT FORM

20 April 2011

The project of **Restoration of Baojiatun Watermill and its Water Conservancy Facilities** is being submitted for consideration for the 2011 UNESCO Asia-Pacific Heritage Awards. Please comment on your experience with this building, such as how well it works, how it relates to the community, etc.

Please return comments to **Prof. Shixing Bao** so it can be included in the entry materials.

The water facilities in Baojiatun played the most important role in the normal life and production of the villagers in the last 600 years. After the restoration project, the facilities still can work properly today, for irrigation, flood control and water storage. It creates a very strong background of landscape architecture and local culture.
In the Marchi of 2010, the extreme droughty weather hit the Guizhou Province. While the other places were suffering from the disaster, Baojiatun village was still covered with green trees and flowers, with a splendid landscape. The restoration project made a great contribution to us all.

鲍家屯水利工程 600 多年来一直是维系鲍家屯人生产生活的生命线。经过水碾房的修复工程，鲍家屯水利工程设施至今仍可以发挥重要的灌溉加工、防洪排涝和蓄水抗旱作用，并为鲍家屯创造了深厚的历史文化和建筑文化景观。
2010 年 3 月初，正是云贵地区百年不遇的大旱之时，鲍家屯仍是青山绿水，景致迷人。水碾房及其水利设施的修复为此作出了重要的贡献。

鲍中任
Zhongren Bao

Baojintun village, Anshun, Guizhou, China

千年火
內外姻親
丑午二王
五谷大神
助筆魁神
親師位
當年太岁
至德真神
灶王

四、鲍家屯水碾房获奖信息及相关研究

UNESCO
Asia-Pacific Awards for
Cultural Heritage Conservation
2011
The restoration of the ancient watermill in the village of Baojiatun sets an outstanding precedent for safeguarding a living agricultural landscape in China. Innovative partnerships among various agencies underpinned a holistic approach to the conservation of the site. The project has returned the watermill, which was the vital linchpin of the water management system dating back over six hundred years, back to working order, along with associated dikes, irrigation channels and dam structures. The highly sensitive restoration work has ensured that the watermill has retained its authentic character and historic patina. In sustaining the functionality of the traditional agricultural complex and associated cultural practices of the local community, the project showcases the significance of Asia's cultural landscapes which are rapidly vanishing under pressures of modernization.
UNESCO
澳門基金會
FUNDAÇÃO MACAU
Award of Excellence
Baojiatun Watermill
Guizhou Province, China
Owner:
Baojiatun Village
Individual Responsible for the Project:
Shixing Bao
Architects/Designers/Consultants:
Shixing Bao, Qingzhou Wu, Xuming Tan, Zhongxing Bao
General Contractors:
Villagers of Baojiatun

联合国教科文组织

2011年亚太文化遗产保护奖

鲍家屯古水碾房的整修项目，树立了中国农业景观保护的卓越范例。参与此项目的各个机构之间的合作与创新，为该古建筑的全面、整体的整修打下了基础。该整修工程使始建于600余年前的水利灌溉系统中的重要环节——水碾房，重新运转起来。它与有关的堤坝、灌溉渠系在一起，使水利灌溉系统恢复了正常的运作。高度认真和细致的整修工作，使水碾房保持了真实的历史面貌，重现了当年的辉煌。在保持传统的农业设施功能和当地文化活动的同时，这一项目还展示了在现代化发展的压力下，保护正在迅速消失的亚洲文化景观的重要意义。

鲍家屯村村委会（使用者）对鲍家屯水碾房及水利设施的使用评价

贵州鲍家屯水碾房是明代军屯制度与地方地域文化特征结合的优秀范例。水碾房和周边的堤围结合，布局合理，构思巧妙，具有结构简单实用，刻意延长坝轴线，尽量利用岩基，就地取材的特点，有灌溉排水、水力加工、供水、游泳、造园、保护生态环境、防洪、军事防御与交通、促进民族团结十大综合功用，使鲍家屯成为“黔中江南”。鲍家屯水利设施有贵州小都江堰之美称，规模虽小，却同样造福当地百姓。水碾房历经600年风雨，一直是维系鲍家屯人生产生活的生命线，使鲍家屯多年地灵人杰，人寿年丰。经过水碾房的修复工程，鲍家屯水利工程设施至今仍可以发挥重要的灌溉加工、防洪排涝和蓄水抗旱作用，并为鲍家屯创造了深厚的历史文化和建筑文化景观。

2010 年、2011年、2013年，云贵地区屡遭百年不遇的大涝大旱，鲍家屯仍是青山绿水，景致迷人。水碾房及其水利设施的修复为此做出了重要的贡献。

中国贵州安顺鲍屯村委会

吴庆洲教授在鲍家屯参加
水碾房文化研讨活动

贵州安顺鲍家屯水碾房修复项目
获联合国教科文组织 2011 年亚太遗产保护奖

贵州安顺鲍家屯水碾房修复项目日前获得联合国教育科学及文化组织 2011 年亚太遗产保护最高奖卓越奖（排名第一）。评委会对鲍家屯水碾房修复项目评价很高，称"鲍家屯古水碾房修复项目，树立了在中国进行农业景观保护的卓越范例。项目高度敏锐而小心地确保复原了水碾房的真实面貌，重现了它的历史光辉。通过对传统农业设施功能的可持续利用以及和当地文化活动的结合，这一项目为我们展现了保护在现代化发展压力下快速消失的亚洲文化景观的重要意义。"

鲍家屯水利工程已经历 600 余年的风雨，重要节点水碾房修复项目由城市规划专家、中国城市科学研究会原副秘书长鲍世行先生主持，采用修旧如旧的原则，完全按照当时的工艺和结构进行修复。鲍世行先生称："鲍家屯水碾房体现古代劳动人民的智慧，把水能变成动能，方便人们生产与生活。它在当地得到'修旧如旧'的保护。作为古代绿色环保水能的利用工程，它的获奖有极其重要的历史文化与科学意义。"

2011 年
6 月
总第十六期

安顺市屯堡文化学会

鲍家屯明代古水利荣获文化遗产保护奖

6月8日，联合国科教文组织在泰国曼谷召开会议时，决定授予中国贵州省安顺市西秀区大西桥镇鲍家屯村明代古水利"文化遗产保护奖"。据悉，联合国科教文组织官员将择日到鲍家屯考察，同时授奖。

鲍家屯明代古水利始建于明代，因为设计科学，布局合理，600 多年来坚持祖先制定的"岁修"制度，并立"禁止碑"严禁破坏，因此现在仍然保持原样并发挥抗旱排洪作用，被誉为"黔中都江堰"。

<北京电讯>

屯堡人

TUNPUREN

2011 年
6 月
总第十六期

黔新出[报刊]2009 准字第 146 号

主办：中共西秀区委宣传部　安顺市屯堡文化学会

鲍家屯明代古水利荣获文化遗产保护奖

屯堡人的民族情结

浅谈屯堡文化生态保护区的构建与旅游开发

黔中早报　2011年9月14日　星期三　编辑：郭妍妍 张璐　美编：向炯　版式：刘芊　校对：玉庆　社会　A Ⅲ 05

大西桥六百岁"水磨"获世界赞誉

联合国教科文组织：大西桥镇鲍家屯水碾房树立了中国古水利利用的典范

本报讯(文/图 记者 梁惠焜 通讯员 吴忠贤) 昨天，记者从西秀区大西桥镇鲍家屯获悉，由北京专家代为向联合国教科文组织申报的2011度亚太地区文化遗产保护奖已经揭晓，鲍家屯水碾房因树立了中国古水利利用的典范，被评为该文化遗产保护奖的最高奖卓越奖。

西秀区大西桥镇鲍家屯是六百多年前明王朝"调北征南"与"调北填南"形成的带有军事防御功能的古村寨。这个有六百多年历史的古村落具有十分鲜明的特色。其中，古水利的利用最为有特色，其引蓄结合的塘坝式工程体系中的水仓坝采用"鱼嘴分流"方法施工，功能、原理跟四川都江堰相似。鲍家屯河上构思精巧的水碾房，是六百多年前鲍家屯人利用水动能舂米的设备。

风景如画的鲍家屯水碾房

600 多年前的古水碾还能工作

鲍家屯完好的古水利吸引省内外众多专家前往考察。中国城市科学研究会首席专家鲍世行等在考察鲍家屯水碾房等古水利利用后评价甚高。今年初，在得知联合国教科文组织开始受理2011年亚太地区文化遗产保护奖项目申报时，鲍世行教授着手收集鲍家屯水碾房相关资料，并代为向联合国教科文组织进行了申报。今年6月至8月，10名国际文化遗产保护专家在泰国曼谷进行评选。9月1日，评选结果揭晓，大西桥镇鲍家屯水碾房获得2011年联合国教科文组织亚太地区文化遗产保护奖最高奖卓越奖。

昨天上午，记者电话采访身在北京的鲍世行教授，年近八旬的老人激动地说："评委会对鲍家屯水碾房评价很高，称它树立了中国古水利利用的典范，是'中国文化遗产保护的生活景观'。"

鲍老称："鲍家屯水碾房体现古代劳动人民的智慧，把水能变成动能，方便人们生产与生活。它在当地得到'修旧如旧'的保护。作为古代绿色环保水能的利用工程，它的获奖有极其重要的历史文化与科学意义。"

鲍老介绍，今年联合国教科文组织亚太地区文化遗产保护奖除了鲍家屯水碾房获得最高奖外，中国还有四川与青海的两处建筑获奖。他已被邀请参加颁奖典礼。不久后，鲍家屯水碾房获奖授牌仪式将在鲍家屯举行。

相关链接：始设于2000年的联合国教科文组织亚太地区文化遗产保护奖初设卓越奖、杰出奖、优秀奖、荣誉奖，2005年增设创新奖。该奖旨在奖励民间相关个人与组织，或与地方政府合作，在保护地方遗产、彰显其文化价值方面所作的贡献。今年该文化遗产保护奖共有34个项目入围，它们分别来自亚太地区10个国家的公共机构、桥梁、酒店、宗教建筑等。

部分相关网络报道

1. 明代乡村水利修复示范工程（2008 年 4 月）

http://wenku.baidu.com/link?url=ZeJFV754uqfbbwel07sXXtaCclpZVgFqvetevqROCkU54YkMa87OyZB5fAO7Bu50PQSv6vEpw2y7U-RDQrkrmDO_AnuVhpcPsLkuBYsr1J7

中国水利水电科学研究院水利史研究所由于一次偶然机会发现了这一完整的古代乡村水利工程。2004 年，通过院专项课题对鲍屯古代水利工程及其价值进行研究；2008 年这一项目得到了国家文物局指南针计划支持，列入该计划的示范项目中。在水科院水利史所主持下，通过与文物保护、古建筑等相关机构或专家、当地水利部门的多方合作，保护项目得以顺利实施。项目对堰坝局部的修复和必要的疏浚，并重点修复了现存唯一的水碾房及其水轮装置。

2. 第五批中国历史文化名村（2010 年 12 月）

http://www.xinhuanet.com/chinanews/2010-12/24/content_21711800.htm

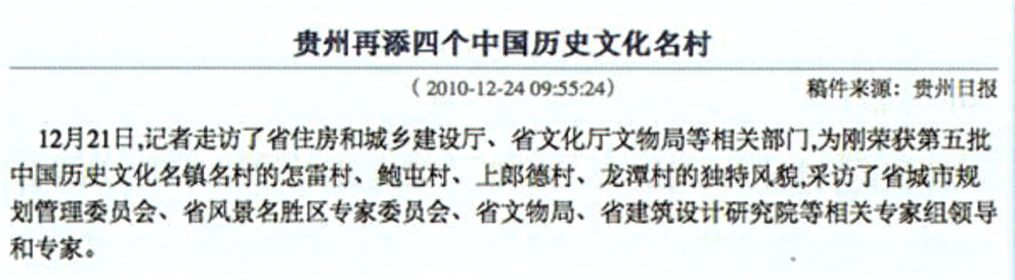

贵州再添四个中国历史文化名村

（2010-12-24 09:55:24）　稿件来源：贵州日报

12月21日，记者走访了省住房和城乡建设厅、省文化厅文物局等相关部门，为刚荣获第五批中国历史文化名镇名村的怎雷村、鲍屯村、上郎德村、龙潭村的独特风貌，采访了省城市规划管理委员会、省风景名胜区专家委员会、省文物局、省建筑设计研究院等相关专家组领导和专家。

3. 全国“三普”百大新发现之一（2011 年 12 月）

http://www.10000xing.cn/x062/2013/0612154555.html

【鲍氏网新闻】鲍家屯水利工程入选全国“三普”百大新发现

中华鲍氏网 2013年6月12日 鲍世行供稿

鲍家屯水利工程入选全国“三普”百大新发现

编者按：据2012年1月13日《贵州日报》报道，鲍家屯水利工程已入选为“全国“三普”百大新发现”，是贵州省三项百大新发现之一。现将该报道转载如下，以飨读者。

百大新发现贵州省有三项

http://www.gog.com.cn 12-01-13 09:55 金黔在线－贵州日报

第三次全国文物普查百大新发现我省有三项在不久前结束的第三次全国文物普查百大新发现的评选中，贵州3项重要新发现从300余项参评项目中脱颖而出，荣膺全国“三普”百大新发现。

这三项新发现分别是锦江谷岸遗址群、鲍家屯水利工程和茅台酒厂历史建筑群，分别属古遗址、古建筑和近现代重要史迹及代表性建筑，是贵州不同历史发展时期的重要实物见证。

---

4. 第七批全国重点文物保护单位（水利）（2013 年 5 月）

http://wenku.baidu.com/link?url=QyuiTrVU6KFeS4pI6lFKlqqXrCcfPQpSUvivJDZdo9LIIl6voe9MFB1cVf-3LLqP962Al548udrnhMGK8tdDJXivbVf-t1XphlXghBVLf4Hy

第七批全国重点文物保护单位名单

（共计 1943 处）

一、古遗址（共计 516 处）

序号	编号	名称	时代	地址
1356	7-1356-3-654	鲍家屯水利工程	明	贵州省安顺市西秀区

佛山兆祥黄公祠修复工程

一、佛山兆祥黄公祠简介

项目基本信息

项目名称：佛山兆祥黄公祠修复工程

工程地点：广东省佛山市禅城区兆祥公园内

完成时间：2002年

占地面积：3060m^2

建筑面积：1712m^2

保护级别：修复前为佛山市级文物保护单位，现为广东省文物保护单位

建设单位：佛山市公用事业管理局

设计单位：华南理工大学东方建筑文化研究所，广东省国际工程设计有限公司

施工单位：五华县第一建筑工程公司

监理单位：佛山市禅建监理公司

历史背景

佛山是岭南中成药的发祥地之一，老城区内有很多与之相关的历史遗迹，兆祥黄公祠即为其一。兆祥黄公祠是著名中成药“黄祥华如意油”的创始人黄大年先生的祠堂，也是佛山老城现存规模最大、型制最完整的祠堂建筑群。该祠堂位于现福宁路与建新东路交叉口的东南方向，处在佛山老城从汾宁古渡到栅下的折线形城市空间发展轴上。祠堂兴建历时15年，从清光绪三十一年（1905年）开始至“民国”九年（1920年）建成。祠堂坐西朝东，占地约3060m^2，建筑部分东西长67.57m，南北宽31.40m。建筑平面布局为三路四进三开间，中轴线上布置头门、拜亭、正厅（祭堂）、寝堂、后楼等主体建筑，两侧对称布置边路，以巷道与主体建筑连接，为广府地区清末民初的典型祠堂形制。祠堂主体建筑除拜亭为歇山顶外均为单檐硬山顶，最后一进建筑为两层高。建筑格局完整，装饰华美，在佛山市的祠堂建筑中极具代表性，具有较高的历史和艺术价值。

产权情况

本建筑建成于1920年，距修复时已有82年历史。建筑原为黄氏家族所有，现由文化部门托管，民间机构租赁。

在80余年的历史变迁之中，黄氏家族的后人基本上移居海外，兆祥黄公祠亦失去了原有功能，长期被闲置，曾经先后作为国民党南海县党部和军队的驻地，自20世纪50年代起祠堂被用作部队营房。“文化大革命”中，祠堂的很多艺术构件遭到严重破坏。1985年，一名部队战士在学习开汽车时撞折了头门的石柱，导致头门屋顶大面积坍塌，先后两度遭遇不合理的加建。但经历

了战乱、社会变革和大规模城市建设活动的冲击之后，仍能幸存并且基本保存完整。1989年，兆祥黄公祠被公布为市级文物保护单位。1992年，祠堂作为危房要被驻军拆除，经过多方努力，部队迁出。1999年，佛山市政府在兆祥黄公祠的周围开辟兆祥公园，并支持修复该祠堂。

修复经过

在民间机构和政府的共同发起之下，对该祠堂的修复工作于2000年正式启动。修复工作采取了多方参与的形式，建立了产权托管人、未来使用者、研究机构、古建专业修缮机构、文化热心人士、政府部门共同参加的多方会议机制。

测绘、研究、修复设计、施工和监理按照严格的技术程序进行，探索了如何利用有限的资金，区别于目前在该地区的古建保护中常见的不合理的虚构装饰方式，采用正确的手段对经历了多次加建、改建的文物进行更有效的保护和维修，重新强调了传统工艺和传统建筑类型的价值，为该地区的文物保护积累经验，也为社区的公共生活作出空间和文化上的贡献。

在多方共同努力之下，整个修复工作共历时2年，于2002年顺利完成建筑修复施工，于2003年交付广东粤剧博物馆正式投入使用，得到了社会各界的关注和新闻媒体的广泛报道，成为运行良好的地方戏剧博物馆。

2000年2月，研究与保护设计启动，建立建设管理部门、文物管理部门、业主代表、研究机构、设计机构参加的多方会议机制。

2000年3月，吴庆洲教授接受委托主持整个修缮的研究和设计工作，确定了“所有修复措施基于对建筑的考古学和历史学研究，基于对历史原真性的保护”的基本原则。

2000年5月，由华南理工大学东方建筑文化研究所完成现场测绘。

2000年6月，进行历史、文化、建筑等方面的研究，进行建筑安全问题的诊断。

2000年5月，根据测绘研究讨论保护与修缮的具体原则、重要技术问题。

2000年6月，提交保护方案，并获得通过。

2000年6月，修复工程开工。

2002年1月，工程竣工验收。

保护宗旨

本保护项目的目的，是在保障兆祥黄公祠结构安全的前提下，通过保存有价值的文化遗产和历史信息，将之既作为以中成药蜚声海内外的黄氏家族乃至整个佛山中成药产业的精神纪念地，也作为城市关键性社会空间，用于展示佛山的传统文化，在今日的城市和社区中继续作为积极的公共场所和文化载体。

经过修缮之后的祠堂用作公共用途，辟为“广东粤剧博物馆”。佛山是粤剧的发源地，该博物馆是全市唯一的粤剧博物馆，所有资金和展品均从社会募集，接受海内外的捐赠，用于展示粤剧的起源、发展和在全球范围内的传播，面向公众开放。

修复的特点与价值

（1）本建筑群的修复由民间机构与政府合作完成。

（2）通过建筑的保存和修缮，保留了佛山本地祠堂建筑的典型范例，以及祠堂作为关键性社会空间的场所精神。

（3）将该建筑与开放空间相结合，成为城市传统空间发展轴上的重要节点。

（4）祠堂的修复大量采用了地方性传统工艺，例如水磨青砖、砖雕、木雕、石雕、灰塑等。

（5）在修缮过程中合理运用了传统技术，例如地方的传统防潮做法、木作和石作工艺，延请传统建筑匠师，并将其技术传授给有志于历史建筑保护的年轻匠师。

（6）建筑材料的选择合理，整体格局和局部样式具有典型的地方性和历史特征：木材用料考究，在不同部位分别使用坤甸木、花梨木、东京木、菠萝格等热带木材。

（7）以其开放性和公共性保证了对周围环境和地方社区的积极贡献，取得了良好的社会效益，并在周边传统建筑已经被大量拆除并建造新建筑的情况下，以其丰富的历史信息和独特的传统特征成为城市的文化地标之一。

修复前的状况

兆祥黄公祠是佛山历史城区内现存规模最大，形制最完整，装饰最丰富的祠堂建筑。但受多次社会变革以及不良使用的影响，建筑多处遭到不同程度的破坏。

1）建筑外观

整体较完整，但头门屋顶大部坍塌，墙身约1/3倒塌、崩裂，走火通道两侧围墙已被拆除，仅余墙基尚存。同时，军队使用期间加建、改建较多，因此墙面上大小洞口多，加建的门、窗户、隔墙和屋顶有多处，地面做法与原做法不符且多处破裂、凹陷，此外，墙面后加的粉刷（石灰砂浆）做工粗糙。室外石条铺砌的地面出现多处凹陷、崩裂。所有建筑的正脊脊饰均已无存，侧墙脊饰除一处之外全部被破坏后抹上灰砂。墙面装饰和木构件装饰被大面积人为破坏，门、满洲窗、格栅窗等分隔和装饰构件几已无存，只留下一些洞口、门臼等痕迹。

2）室内

拜亭下的花砖地面已产生明显的隆起。室内原有的匾额、楹联、屏风等小品已全部佚失，仅剩挂钩若干。后进天井石阶、墙身均有小孔，估计为石栏杆或者石装饰构件。厢房梁有长挂钩，可能为悬挂灯笼之用。室内现无家具、牌位等。

3）结构现状

除头门之外，整体结构受损情况不严重。但是有局部梁、柱遭受较严重的蚁害，出现较大的裂缝，需要加强，屋檐部分桷子糟朽现象比较普遍。有3处墙身出现轻微的变形或裂缝，影响结构安全。人在阁楼上走动时有明显的震动感，各处木楼梯均遭到不同程度的损坏，已不便上人。

4）技术设备系统

供水：南侧走廊西端有眼小井，现已干涸。

排水：原雨水排放系统较完善。屋面多采用直接排水于天井、露天廊的方式。第二、三进天井屋檐下有天沟排水铁构件残留，巷中有多处铸铁雨水管和铁钩残留。各天井、巷中均有铁制或石制雨水集水口。

修复前的损坏情况

1）总体布局

除原有走火通道两侧的围墙被拆除外，整个祠堂总体布局基本完整，保持了原有建筑的风貌。从平面布置和造型的差别来看，后进建造年代明显晚于其余部分，从二进主体建筑进入后进必须经过侧巷，后进建筑的工艺相对粗糙，装饰也相对简单，细节上显示受西方装饰风格的影响更强。山墙没有高起的侧脊。

整个祠堂共4进，头门遭到严重破坏，除二进主体建筑祭堂（C2，编号对应位置详见编号图，以下同）后墙被封堵、三进主体建筑寝堂（E2）后墙被多开了2个洞口之外，各主体建筑基本保存原有格局，但两侧厢被改建较多。多个小院、天井被覆盖，通向户外的门被封堵，且存在大量的加开门窗洞口的现象。Y3通向B3的门被封堵。

2）屋面

因人为破坏和长时间未检修，屋面普遍存在破损、糟朽、漏雨等现象。从现状看，屋面的各种损坏类型包括：

（1）头门处屋顶大面积坍塌；

（2）巷道上有多处加建的小屋顶；

（3）多处屋顶存在漏雨现象，比较集中的是头门两侧厢（A1、A3）、拜亭（B2）、二进主体建筑（E2）。

（4）糟朽，尤以各次要建筑为甚。因为经雨水浸泡较多且易受风的影响，屋面下部的瓦垄灰脊、瓦面普遍已经松动，飞桷普遍存在糟朽现象，已经成为安全隐患。

（5）屋顶被改建。头进南侧厢（C1）斜屋面和卷棚交接处屋面被改建，改建处存在多处不合理的构造，施工使用了现代材料和现代做法，手工粗糙，与原有建筑风格明显不符。

屋顶装饰被破坏。所有建筑正脊均被破坏；头进院落两侧建筑（B1、B3）飞子与椽条交接处的压脊仅有局部残留；后进F1、F2屋顶原有装饰构件被破坏，从现状遗留来看，该构件应为灰、瓦塑成，但形状、内容已不可考。

3）墙面

墙面的主要损坏类型包括：

（1）头门处大面积坍塌；

（2）墙身变形，较明显的有3处，A3东侧墙身顶端有宽约10cm的裂缝，南侧墙身有明显倾斜，C3西侧墙身中央部分向外突出；

（3）部队使用期间多处添建了青砖、红砖、木板隔墙，红砖隔墙现已基本拆除；

（4）部队使用期间多开了大量门窗洞口；

（5）部队加建的阁楼、偏厦拆除后在墙上留下大量直径约25cm的小洞口；

（6）原有门窗洞口被堵塞，主要是现在朝外的各边厢天井侧门；

（7）墙面被加以石灰批荡；

（8）添建了大量门楣和窗楣。

4）地面

现有的室内地面铺装共4种：各阁楼处（G2、C1、C3、A1、A3）为斜45°铺砌的灰色尺二方砖（360mm × 360mm × 60mm），拜亭处为20mm × 20mm的花砖，后进F1、F2地面为红色尺四阶砖（470mm × 470mm × 60mm）错缝铺砌。阁楼地面为最初铺装，对缝精细，保存完好，无须另外处理。厅堂地面铺装为后加，勾缝且手工较粗糙，地面多处凹陷、崩裂，砖下沙层已有较多陷落、流失现象，而且加建隔墙的墙脚仍存，所以需要对厅堂地面按照最初做法重做垫层，补充河沙，重铺方砖。

各天井、院落、内廊及头门外平台地面均为长条青石铺地。受多次施工影响，已有多处产生凹陷及崩裂。这些凹陷除影响观瞻和交通外，同时不利于水的排泄，易在院落和巷道中产生积水。此外，祠堂背后地面凹陷严重，需要加高土垫层至西墙石条底部。

5）木阁楼

祠堂内现有木阁楼7处，分别位于A1、A3、C1、C3、G1、G2、G3内，其中G1、G3为新中国成立后添建，痕迹明显，用料粗糙，规格不一，工艺简单。原建阁楼结构部分为整齐、均匀、做工精致的矩形密肋，阁楼表面为拼缝的360mm × 360mm的灰砖贴面。加建部分用材混乱，施工粗糙，现已局部拆除，建议全部拆除。

C1阁楼明显被改建，楼梯的平面位置不合理。

6）柱

结构柱包括木柱和石柱两种，其中木柱共计22根，石柱共计14根。

木柱现状基本良好，但底部普遍有裂缝，有多根遭到不同程度的白蚁蛀蚀，其中有2根相对严重。柱身上普遍有拆除加建部分之后的残留物。

石柱除头门处被撞断的一根之外，其余基本完好，但柱础角部细小的崩缺较多。加建的砖隔墙拆除之后，在大量石柱上留下了石灰痕迹，并有部分未清理干净的砖角，头门靠内的2根石柱被加建的墙体所包。

7）梁架

梁架材料为坤甸木，除头门梁架被严重破坏外，现状保存基本完好。局部遭受蚁害，并有部分被改建过，局部增加了檩子。在G2南次间梁架有加建的支撑。

受天花遮挡，难以仔细察看G2处加建支撑的原因，但未发现有严重的损坏。

8）木构件装饰、门窗

挂落均已无存，根据修复原则，保留其痕迹，不再复原。

木雀替共缺失5个，均位于C2外侧梁架。

兆祥祠所用木斗栱不多，除头门外，梁架部分斗栱保存基本完好。

封檐板基本被保留，但其上装饰受到不同程度的破坏。

E1、E3格栅窗有局部留存，其余格栅窗基本只余边框。

除入口大门外，祠堂所有满洲窗、木门均已佚。

9）石构件装饰

石梁、驼墩、栏杆、望柱基本保存完好，有几处缺失被补上水刷石构件。局部有破损、移位现象，或被砖墙埋没。

建议拆除多余砖墙后清理石构件。缺失部分不再修复。

10）其他装饰

正脊缺失，装饰在“破四旧”时遭受了集中的破坏。

11）排水系统

建筑原有的排水系统包括铁制天沟、铁制水落管、雨水收集口（石制、铁制两种）地面明沟、地下排水沟，设计较完善，但是未能得到有效保护，因为排水系统仍将发挥功能作用，建议恢复所有的排水构件，整饬排水坡度，疏浚排水管、沟。

12）防火

在公园设计中未预留消防车道，从现状来看，祠堂四周已经被绿化和公园小品占满，难以开辟环行或者沿长方向的消防车道。鉴于本建筑群基本为单层建筑，建筑长度仅为70m左右，且四周均较开阔，所以不设置消防车道，采用车用消防栓、水带用消防栓和灭火器相结合的消防系统。

在祠堂靠近建新路一侧（北侧）设置2个车用消防栓，分别位于东、西两端；在南侧设置2个水带用消防栓；在祠堂各院落内均布置2个泡沫灭火器，置于角落处木架上，不使用悬挂方式以免破坏建筑的墙面。

13）防雷

未发现原有的防雷设施。

14）防蚁

遭到白蚁破坏的构件比较集中在C2、E2：

（1）C2正背下往东第三条有白蚁，裂纹较大，需加固处理；

（2）C2明间正背下往东第四条有白蚁，声音有空洞；

（3）C2从南梁架向北第七条瓦沟有蚁路；

（4）从C2北墙向南第三、六、七、八、九、十一、十三、十五条瓦沟有蚁路；

（5）C2东北柱有蚁。

修复原则

（1）首要的指导原则：在修缮的过程中保持文物的原真性。不添加构件种类和不可移动的设施，各种技术、安全措施均不改变文物的原貌。

（2）经审慎研究后清理、拆除后人不合理的添建、改建，保持原建筑的风格。本原则适用于部队使用期间加建的门窗、墙、屋顶、阁楼等。这些加建部分破坏了祠堂的本来建筑面貌，损害了祠堂的整体性和艺术价值、历史价值。因为加建、改建之处较多，在审慎地进行辨别、研究，根据材料、工艺、装饰风格、残留的细小痕迹等确定被改动的部位之后，拆除多余的不合理的添建、改建。

（3）关于建筑材料的原则：对明显可考证且易于修复、恢复的部分或者影响建筑物正常使用、意义重要的部分进行整修、恢复，尽量保持原有构件与材料，修旧如故；如必须更换则采用相同材料，原构件无法再用的补以新构件，修补复原，并与旧材料进行区分。

（4）无法考证且不影响建筑物正常使用的缺失构件不进行复建。部分装饰构件已经全部佚失，虽然其尺寸、大致轮廓可以根据残留的痕迹和同时期本地祠堂建筑装饰的一般风格大概推断，但其真实面貌已经无法考证，因为不影响建筑物的正常使用，依照本原则，不予重修，例如，现已全部被破坏的正脊脊饰不再补回。但是对于其残留的痕迹和构件则予以仔细保留。

（5）排危和延长建筑物寿命的原则：消除或尽量减少各种安全隐患，在尽量不影响外观的情况下，对祠堂的结构、排水系统、防火措施、防雷等技术内容进行全方位的检查和改进，尽可能避免灾害的发生，在万一发生灾害时使其保持一定的应变和抵抗能力。所有因为技术改进而增加的构件（例如结构支撑、消防栓等）尽量独立在原建筑之外，并且视觉上可以明显区别于原有建筑。

（6）对所有房间、院落、结构构件、门窗和其他独立构件进行清晰、准确的技术编号，建立细致的技术档案。

项目的复合性、敏感性和技术一致性

本项目的目标不仅仅是为了消除历史遗产的不安全因素，而且是为了使历史遗产获得适当的开放性和公共性，在持续的使用中对周边环境和社区作出积极的影响，延续传统的精神。这决定了本项目方法论在以下几个方面关注复合性、敏感性和技术一致性的问题。

1. 复合性

保护工程在消除不安全因素，恢复建筑物原真状态的同时致力于创造适当的开放空间和公共空间。此项任务贯彻于项目的全过程。

2. 敏感性

为了满足现代社区活动需要的空间，环境设计在以下几个方面充分考虑了历史遗产对于工程措施的敏感性：

（1）空间尺度的历史感

建筑的使用计划是作为小型的社区博物馆，在合理安排展品和参观流线的前提下，遗产现状的空间布局可以实现这一新的使用功能，因此本项目强调了对建筑物现状的充分尊重。

（2）装饰物和设施的环境感

对于现代使用功能不可避免的新增设施（例如展览、照明、消防等），本项目要求在降低其对于历史环境的视觉冲击性的同时，保留必要的可识别性。

本项目强调不在建筑遗产中增加任何新的装饰物。对于为社区博物馆的使用目标而增添的少数具有展示价值的装饰物必须是可以移动的，不得固定于建筑物之上。

（3）环境影响

防止大气、粉尘污染：对于现场的木机械和青砖切割机械，经常洒水湿润，减少粉尘污染。建筑施工垃圾采用容器吊运，及时清运，适量洒水，减少扬尘。

防止水污染：凡进行水作业时产生的污水，控制好污水流向，防止蔓延。施工污水严禁流出工地污染环境。对现场存放油料的库房进行防渗漏处理。

防止施工噪声污染：施工现场遵守《中华人民共和国建筑施工现场噪声限值》(GBJ 2523—90)，制定降噪的相应制度和措施，严格制定作业时间，采取降噪措施，做好周围群众工作，报有关环保单位备案后方进行施工。

修复技术措施

（1）加固：结构加固、构件修补、构造完善；

（2）复建：恢复了部分缺失或损坏的梁架。对部分遗失且在结构或功能上必需的构件进行了复建，对于无法考证原始面貌又不影响结构安全的装饰部分，不进行复建。根据博物馆的展出需要，增加了室内陈设。

（3）新旧关系：新增加的陈设均可移动，其风格与原来的建筑相协调，又有可辨识的区别，添加的部分遵循“可逆性”原则。

具体修复措施：

1）总体格局

鉴于祠堂四周已经按照公园的统一设计布置了许多已完成的建筑小品，而且原走火通道围墙的复建不存在功能上的必要性，故建议该围墙不再复建。

建议恢复祠堂内部的建筑格局，去除不合理的添建和改建，恢复应有的通道、天井。各朝向户外的门建议去掉现有的封堵物，加门扇以方便管理。

2）结构安全与正常使用的需要

（1）屋面木构与瓦作

头门大面积坍塌的屋顶予以复原，重铺瓦面。

拆除巷道中加建的屋顶。

对于结构状况较好但存在漏雨现象的头门两侧厢（A1、A3）、拜亭（B2）、寝堂（E2）等处屋顶进行局部修缮。

对其他屋面的檐口部分全部进行结构构件更换和瓦面、滴水、瓦当的重铺。

（2）墙面砖作与门窗

头门被损坏的墙体部分全部按照原做法复原，外墙面全部采用磨砖对缝，头门北侧厢（A3）东墙从裂缝开始部位起重做墙身，外墙面磨砖对缝，所用青砖全部采用原有材料，不足部分从民间收集，新磨的青砖色彩与原有青砖可辨别。受附近住宅建筑打桩的影响，A3东墙产生裂缝，南墙产生裂缝、倾斜变形，已经对屋面产生了拉动导致多处漏雨，而且在结构安全上有一定的隐患，对东墙、南墙从裂缝、倾斜开始的部位起重做砖墙。

去除所有现有的墙面批荡，而且不再另加批荡。因为各进主体建筑墙面均为青砖磨砖对缝，次要部分为淌白做法，外观效果、物理性能良好，另加批荡反而会破坏建筑的本来风格，所以小心去除现有批荡之后，对墙面进行清洗，还墙面以本来面目。

对各种洞口进行封堵，按照原来做法还原。原有门窗做法统一，门、窗的各边全部有花岗石条，宽度11cm；门楣、窗楣做法讲究，装饰细致；窗户的铁栏全部采用直径30mm的圆铁，门槛为石条，门臼、转交做法精细，且门窗上方经常有灰塑或者变化的砌砖方式等装饰。后加的门、窗户均无石条边框，边梃、上梃均为木制，窗扇分格采用的是晚近的做法，铁栏钢筋较细，门楣、窗楣为木板加抹灰，装饰简单。由此判断，兆祥祠原有的窗户共20个，巷道、走火通道上的小门（900~1200mm）共22处，其余均为加建，对所有有门臼的门洞，补回门扇。

拆除多余的墙和隔板。头门的内部隔墙、西侧墙面均为加建。

（3）地面铺砌

根据传统做法，屋顶坡向为南北方向的次要房间采用尺二方砖错缝密铺，不勾缝；屋顶坡向为东西向的主要房间采用尺四方砖错缝拼缝密铺。拜亭处花砖为祠堂原物，但中部凹陷严重且有几处破损，使用原物重铺，换铺垫层，以便排水。

各天井、院落、内廊及头门外平台地面均为长条青石铺地。受多次施工影响，已有多处产生凹陷及崩裂。这些凹陷除影响观瞻和交通外，同时不利于水的排泄，易在院落和巷道中产生积水。对所有凹陷、崩裂的部分进行重铺，巷道上的石条应靠近墙角的部分低，巷道中间高，使地面的自然排水得以实现。

（4）木阁楼

拆除G1、G3的加建部分。

C1对照C3的平面复原。后进G2靠西墙部分原为镂空，从一层看天花、阁楼看地面拼缝可以判断加建的部位，拆除加建部分，在镂空的地板周围加建栏杆。

原有阁楼上人时均有明显的震颤感，这不是因阁楼结构部分年久失修引起的，相反，肋、木楼板和灰砖表面均保存完好，并无腐烂或者断裂迹象。震颤感的原因是木材本身的特性造成的，在跨度较大的情况下，木板的跨中挠度较大，故在跨中感受到的震颤感最明显，即使全部拆除新建，在不改变跨度的情况下，仍然不会有实质性的改进。建议不拆除，修复楼梯、栏杆，限制使用。

多处阁楼上有后来加建的天花板，建议全部拆除。

（5）柱

木柱现状基本良好，但底部普遍有裂缝，有多根遭到不同程度的白蚁蛀蚀，其中有2根相对严重。柱身上普遍有拆除加建部分之后的残留物，对于柱身上的多余残留物应小心辨别，祠堂原有的挂落、格栅、门框、门槛等被破坏后的遗痕应小心保留，而铁钉、接线板、木条等应予拆除。

对损伤最严重的一根（MZ14）在底部加铁箍，对所有裂缝采用灌注环氧树脂的处理方式。环氧树脂虽然不会增强承载力，但是可以阻止进一步的蚁害和其他细菌、水分的侵蚀。

石柱除头门处被撞断的一根之外，其余基本完好，照原有样式重凿一根，柱础使用原柱础。其余石柱拆除，清理多余物，清除石灰痕迹。

（6）梁架

建议头门东侧采用原有材料，构件全部落架重修，以保证施工和结构的安全。现门外尚未倒塌的部分亦相当危险，仅依赖倾斜的斗底夹在梁架和檩子之间维持结构稳定。所以必须全部落架，方能保证施工和结构的安全。

增加了檩子的部分将多余构件拆除，同时更换原构件，保持材料和色彩的一致。对于遭受蚁害较严重的最后一进建筑的主梁和部分檩子，分别采用樟木大梁和杉木檩子进行更换。

建议拆除掉G2内加建的临时支撑，但施工前需要仔细观察。

3）材料的选择、获取

被拆除的部分主要采用原来的构件，对照对称一侧的梁架进行编号、组合，且全部采用榫卯连接；不足的部分从民间采集同样的构件，或者使用同样的材料，按照原来的样式重新制作。

4）样式、建造及装饰工艺

建造和装饰严格遵循传统工艺的做法。

（1）木构件装饰

兆祥黄公祠除头门坍塌处外，梁架部分斗栱保存基本完好。因为头门梁架发挥着重要的结构和精神功能作用，头门斗栱全部按照原样恢复。对于“文化大革命”中被斧凿的木雕，保持被破坏之后的情况，以体现不同历史时期的痕迹。

对已经缺失的挂落不再复原，保留其痕迹。

花梨木大门门神痕迹可循，建议修复。头门梁架木雕破坏严重且工艺复杂。

整个建筑群共缺失木雀替5个，因为雀替起到了辅助性的结构作用，而且相对的雀替均完全对称，故按照相对一侧的形状进行复原。

封檐板基本被保留，但其上装饰受到不同程度的破坏，补回缺失的封檐板，因工艺复杂，对板上已被破坏的装饰不再修复。

建议恢复E1、E3的装饰，式样同残留局部。其余不可考的装饰不再恢复。

根据地面和墙面痕迹恢复门扇，但装饰不再复原。格栅窗有局部留存，按照残留局部的式样恢复。

花梨木大门门神循旧迹重描色彩。头门梁架木雕因破坏严重、工艺复杂且可作为建筑所经历的历史阶段的遗痕，故保存现状，不另外修复。

（2）石构件装饰

拆除多余砖墙后清理石构件。缺失部分不再修复。

（3）其他装饰

正脊不再恢复，脊尾饰按照现保留的一个完整的脊尾修复，草龙因为破坏较严重且影响观赏进行修复。小红山色彩还原，灰塑、墀头被破坏处不再修复。室内挂钩保留，用以悬挂楹联、匾额。门窗檐饰、门楣窗楣因为确定可考且工艺清晰，原有门楣、窗楣均按原样修复，后来增建的门楣、窗楣拆除。

5）技术设施

建筑原有的排水系统包括铁制天沟、铁制水落管、雨水收集口（石制、铁制两种）地面明沟、地下排水沟，设计较完善，但是未能得到有效保护，因为排水系统仍将发挥功能作用，修缮工程恢复所有排水构件，整饬排水坡度，疏浚排水管、沟。

6）灾害防治

（1）防火

在公园设计中未预留消防车道，鉴于本建筑群基本为单层建筑，建筑长度仅为70m左右，且四周均较开阔，故采用车用消防栓、水带用消防栓和灭火器相结合的消防系统。在祠堂北侧设置2个车用消防栓，在南侧设置2个水带用消防栓；在祠堂各院落内均布置2个泡沫灭火器，置于角落处木架上，不使用悬挂方式以免破坏建筑的墙面。

（2）防雷

为保护木结构建筑，本次修复采用接地式防雷。为防止破坏建筑的原来形象，在主要建筑的正脊上铺设铁片，上覆水泥抹灰。再用钢筋沿侧脊内侧敷设，沿西侧外墙接入地下。

（3）防蚁

经过开平市白蚁所的处理，兆祥祠曾经比较严重的白蚁灾害现在已经得到了有效控制，初测时尚有主要建筑的几处柱、梁有新的蚁路，在多次补测、复核时发现均已得到处理，未发现新的白蚁痕迹。佛山公用事业管理局已经和白蚁防治所达成协议，定期（1月1次）对兆祥祠进行白蚁状况检查。

（4）其他灾害防治

因四周有较多高大建筑物遮挡，且祠堂屋面突出物不多，不用考虑专门的防风措施。因本建筑群主要为单层砖木结构，结构状况改善之后，亦不用采取专门的防震措施。

修复中的匠艺

1）建筑匠艺

水磨青砖、门窗、防雷等，其中水磨青砖工艺要求十分精细，每块青砖都要磨5个边，经过切块—修边—粗磨—开槽—细磨等多道工艺，最后水磨青砖墙的墙面垂直度控制在了3mm以内。

墙面砌法：空斗墙。陶瓦屋面瓦龙的直顺度控制在8mm以内，瓦面平整度控制在25mm以内。

2）工艺手段

彩绘、石雕、木雕，以及油漆的配比等。

3）修复技术使用铰链和滑轮的起重，防蚁措施，纠偏，防腐处理。

4）修复工程中使用的材料

（1）青砖，用于墙面砌筑和装饰，规格：253mm × 110mm × 57mm（承重墙部分）；253mm × 110mm × 53mm（磨砖对缝部分）。

（2）花岗石石，规格350mm宽，用于室外地面的铺砌；局部的人物石雕采用青石。

（3）大阶砖，包括两种规格：470mm × 470mm × 10mm，红色，用于一层室内地面的铺砌；360mm x 360mm x 6mm，灰色，斜铺，用于楼面铺砌。

（4）彩色水泥花砖，规格200mm × 200mm × 5mm，用于拜亭地面铺砌。

（5）瓦：陶制板瓦、筒瓦、瓦当、滴水。

（6）木材：木柱、斗栱、檩子采用坤甸木，雀替采用菠萝格，挂落及木雕采用柚木，大门采用花梨木，需要增加的构件材料从旧货市场中寻找。

新增元素与技术措施的合理性

1）新添元素

部分用料、重新涂刷的油漆。

2）技术措施的合理性

本次修缮不添加构件种类，各种技术、安全措施均不改变文物的原貌，脚手架不直接接触建筑物，消防器材另设支架而不固定在墙面上，拆除多余构件采用不扩大破坏面积的方式等。

3）施工技术

对规格为250mm × 250mm × 6000mm的石柱，采用井字木架配合2个手拉葫芦的方法，一次吊装归位成功。

磨砖对缝的工艺流程：切块、修边、粗磨、开槽、细磨。因为整个祠堂一共有5种规格的青砖，加工好的青砖按照规格摆放整齐，施工时“对号入座”，墙面丝缝严密，表面的平整度、垂直度均控制在3mm以内。

黄公祠的屋面最多的地方共有7层，落架检修时注意其搭接方式，并作文字记录和拍照，垂铺时注意底瓦、盖瓦的搭接要求，保证瓦垄齐直、平整，裹垄灰粘结牢固，无起泡、翘边、裂缝、露麻等现象，瓦垄直顺度控制在8mm以内，瓦面平整度控制在25mm以内。

后进3根长达6m的屋脊大梁的拆换采取搭木架结合手拉葫芦，第一步先将桁条调至棚面，第二步再将桁条吊至屋顶榫接安放。

4）施工方式对环境的影响

注意控制好噪声、污水、腐蚀性等。

此外，在对建筑物的趣味、环境、布局和安全没有影响的前提下，适当采用新技术。

5）技术一致性

保护工程在消除不安全因素，恢复建筑物原真状态的措施中，强调了技术手段的一致性，即采用传统的材料、传统的工艺来完成所有修缮工作。

采用了“技术交底在前、施工在后、样板引路”的施工程序

尤其选用酯胶磁漆和醇酸磁漆，其中主调采用朱红：黑＝9：7的比例。

主要参与人员与机构

（1）研究与设计人员：吴庆洲、冯树、冯江、欧捷、肖旻。所属机构：华南理工大学东方建筑文化研究所，负责测绘、文献研究、考据、工艺研究、技术研究，制定维修原则，主持修复设计。对建筑中的传统绘画、雕刻艺术进行研究，提供修复工作的学术支持。广东省国际工程设计有限公司配合完成施工图设计。

测绘参与人员：吴庆洲、冯树、冯江、欧捷、周毅刚、肖旻、袁忠、蔡效军、苏畅、刘虹、简言、肖昕、黄骏、林燕、刘穗杰、潘杰威、李敏稚、郑洲、郑嘉颖、杨力研、刘祥海。

（2）主要工匠：谢伟光、邓吉彩。所属机构：广东省五华第一建筑公司。

（3）民间组织：广东粤剧博物馆。一个通过民间渠道筹集资金和展品的博物馆，修缮的主要资金提供者，以及修缮完成之后的租赁者和使用者，并另外提供展品和资金进行祠堂的日常管理和维护。

（4）政府机构：佛山市公用事业管理局、佛山市文化局。政府有关职能部门行使托管权，负责邀请专家、专业人员进行测绘、调查和修复设计，并依照法律程序进行施工招标工作，维修组织、管理和考评等，并承担部分费用。佛山市同时负责协调博物馆的研究、使用和管理，教育当

地民众如何保护和维修地方文物。

（5）黄祥华先生后人。

对用途与维护的规定

（1）对用途的规定：只能用于公共的文化用途，参观、展览，不得用于商业目的。

（2）一次性进入的人数限制：阁楼少于10人。

（3）年度使用人数的限制：不多于20万。

（4）维护的规定：明确日常维护的频率与工作内容，如定期（1月1次）对兆祥祠进行白蚁状况检查，由佛山市文物管理部门进行监管。

修复后的状况

兆祥黄公祠现位于对市民免费开放的兆祥公园内，东面现有一小型广场和方形水池；南侧为公园游廊和比较宽阔的绿地；北侧距建新东路最近点约30m，中间为绿化带，已种有2棵大榕树及数棵蒲葵，疏密间距得当，从建新东路可若隐若现地看到古朴的黄公祠，景观效果良好；西侧距福宁路20m，亦以绿化相隔；在福宁路西距黄公祠背面约50m处有8层的灰色住宅楼和与之相连的2栋白色较新的9层综合楼；其余的高大建筑均隔路相望。

兆祥黄公祠原来作为祠堂使用，修复完成后作为粤剧博物馆使用。因为黄氏家族的后人均已移居香港、澳门和海外，祠堂基本处于废弃、闲置的状态，难以发挥其最初功能。故在不改变建筑形制和外观，保证结构安全以及不在建筑上添加任何不可移动设施的前提下，将原来的祠堂用于公共用途，用于展示佛山良好文化传统的公共建筑使用，且该功能不对建筑产生物理和生物破坏。该祠堂作为粤剧博物馆使用情况良好，2003年6月至2004年12月共接待访问者15万人。

修复带来的影响

（1）通过恰当的保护，兆祥黄公祠在修复完成后，经过专家的评定，2002年7月被列为广东省文物保护单位。

（2）对比于之前完成的一系列文物建筑和历史地段的保护工作，兆祥黄公祠的保护为当地在古建筑保护方面提供了新的思路，促进了佛山市文物部门和规划建设管理部门对传统建筑的地方性和原真性的追求，并将相应的要求列入其后编制的《佛山历史文化名城保护规划》和《佛山老城历史文化核心区保护详细规划》。

（3）兆祥黄公祠保护工程完成后，与周边免费开放的公园一起成了城市中活跃的公共空间节点，目前已经成为周边社区最有活力的场所之一，促使佛山市政府将更多的历史建筑及其周边改造为城市的重要地标——不仅在传统建筑形式的独特性上，而且在开放空间的社会性上。

（4）目前佛山市政府正在酝酿以兆祥黄公祠为开端的福贤路历史地段的保护与更新。

（5）兆祥黄公祠的保护工程成功地实现了政府与民间的合作，这为其他的类似实践提供了经验。

二、工程设计图纸（部分）

现状图

现状首层平面图

现状阁楼层平面图

现状屋顶平面图

现状正立面

现状北立面

现状 1-1 剖面

房间院落编号

地面、柱、门窗编号

修复设计图纸

兆祥公园总平面图

首层平面图

阁楼层平面图

屋顶仰视平面图

正立面修复设计图

背立面修复设计图

北立面

南立面

1-1 剖面

2-2 剖面

3-3 剖面

排水系统平面图

门窗大样图

柱础大样图

三进厅内挂落

封檐板大样（局部）

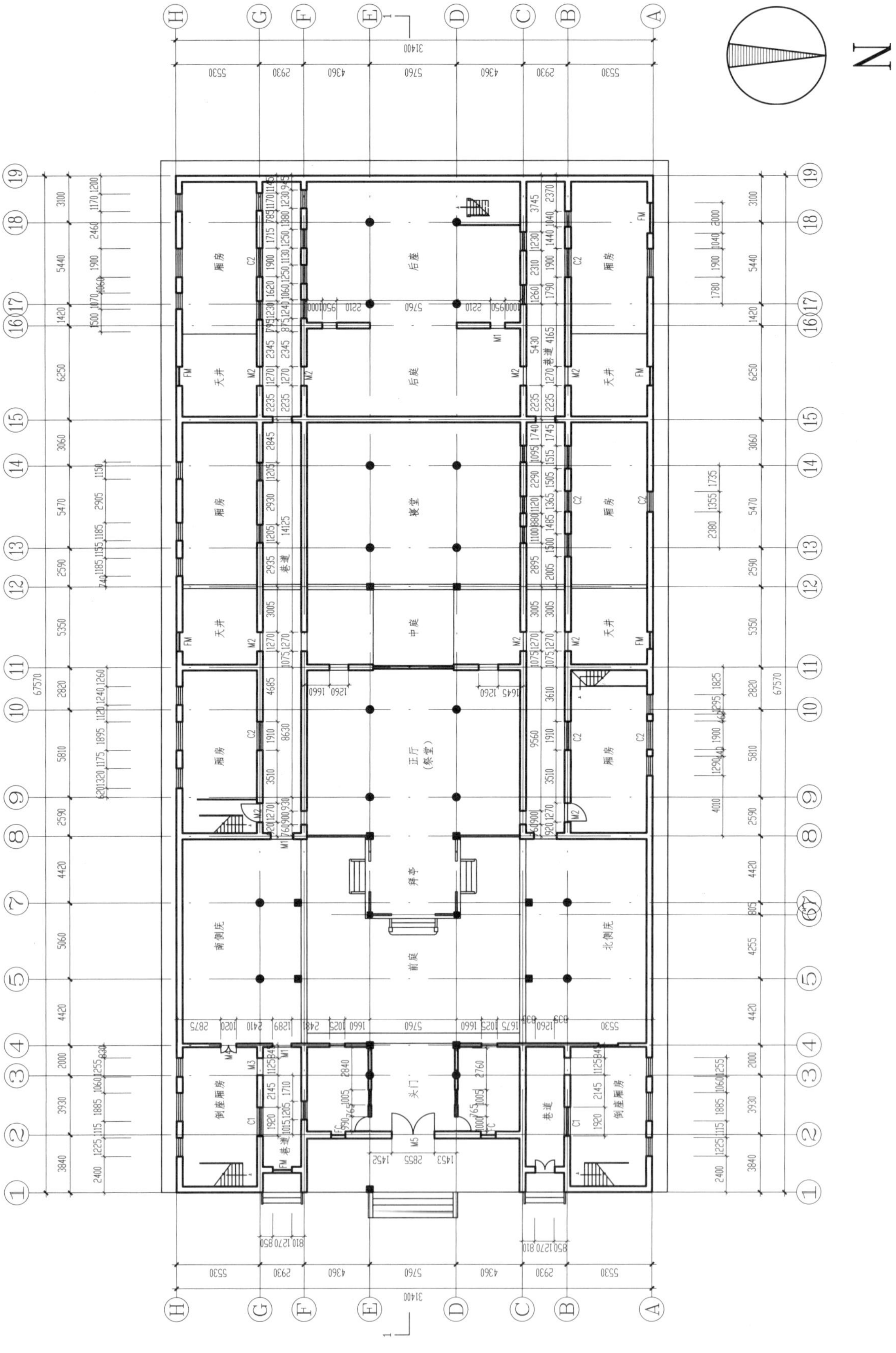

现状首层平面图

现状图

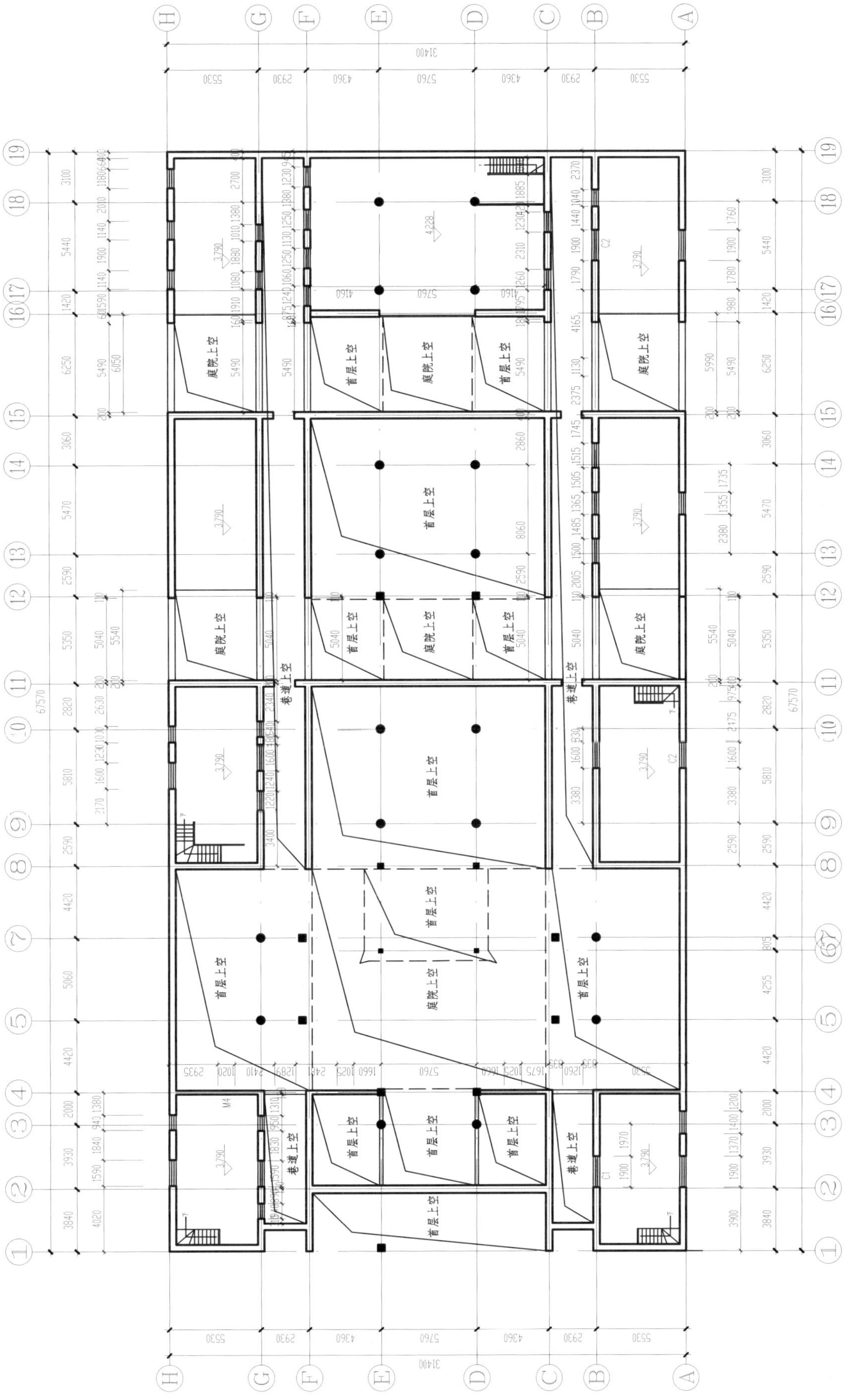
H
G
F
E
D
C
B
A
31400
5530
2930
4360
5760
67570
首层上空
庭院上空
巷道上空
3.790
4.228

现状阁楼层平面图

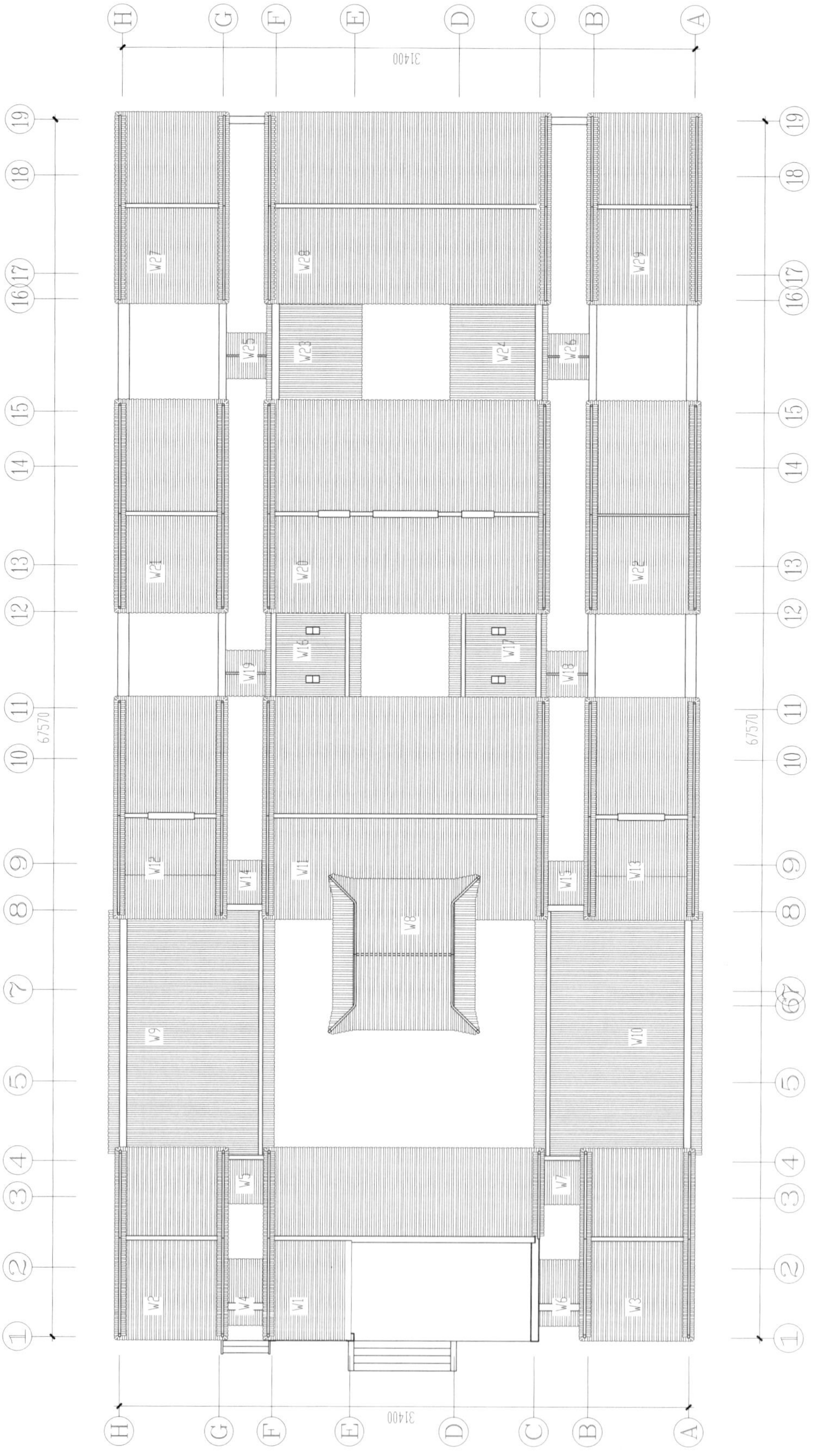
H G F E D C B A
31400
19 18 17 16 15 14 13 12 11 10 9 8 7 6 5 4 3 2 1
67570
W1 W2 W3 W4 W5 W6 W7 W8 W9 W10 W11 W12 W13 W14 W15
W16 W17 W18 W19 W20 W21 W22 W23 W24 W25 W26 W27 W28 W29

现状屋顶平面图

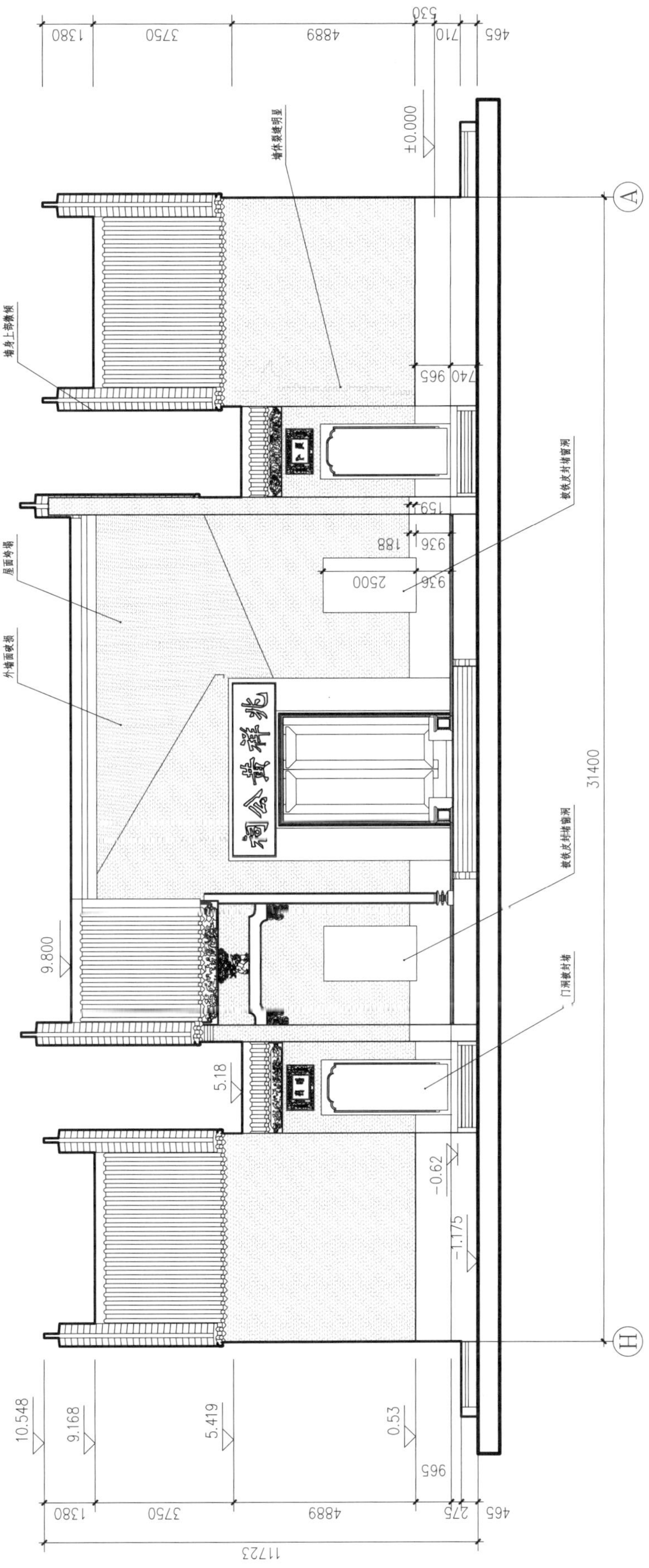
兆祥黄公祠
1380
3750
4889
530
710
465
965
740
159
188
936
2500
±0.000
31400
9.800
5.18
-0.62
-1.175
10.548
9.168
5.419
0.53
275
11723
A
H

现状正立面

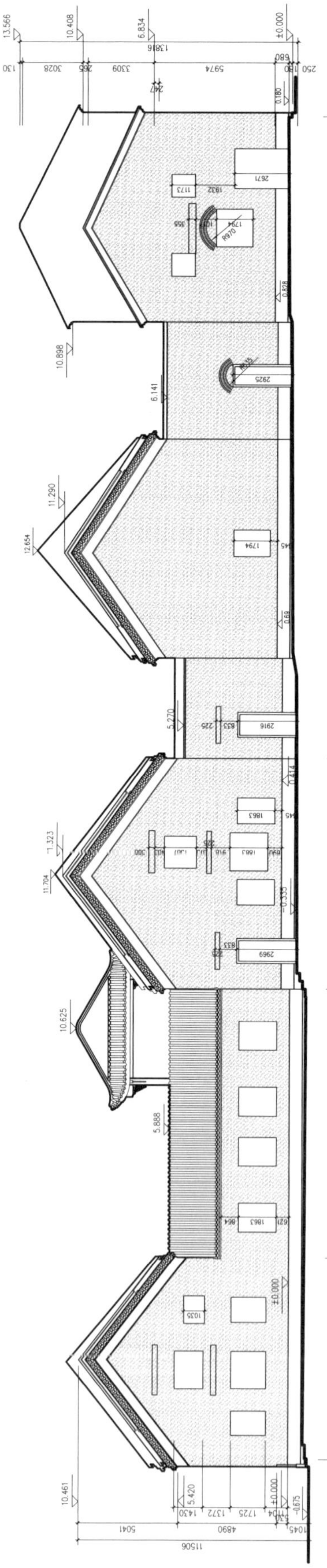

现状北立面

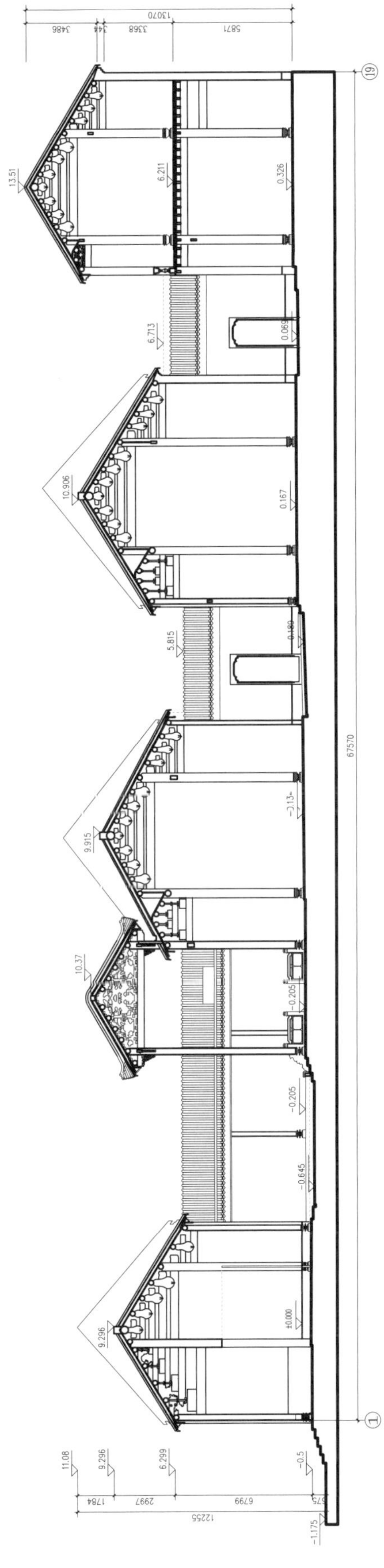

现状 1-1 剖面

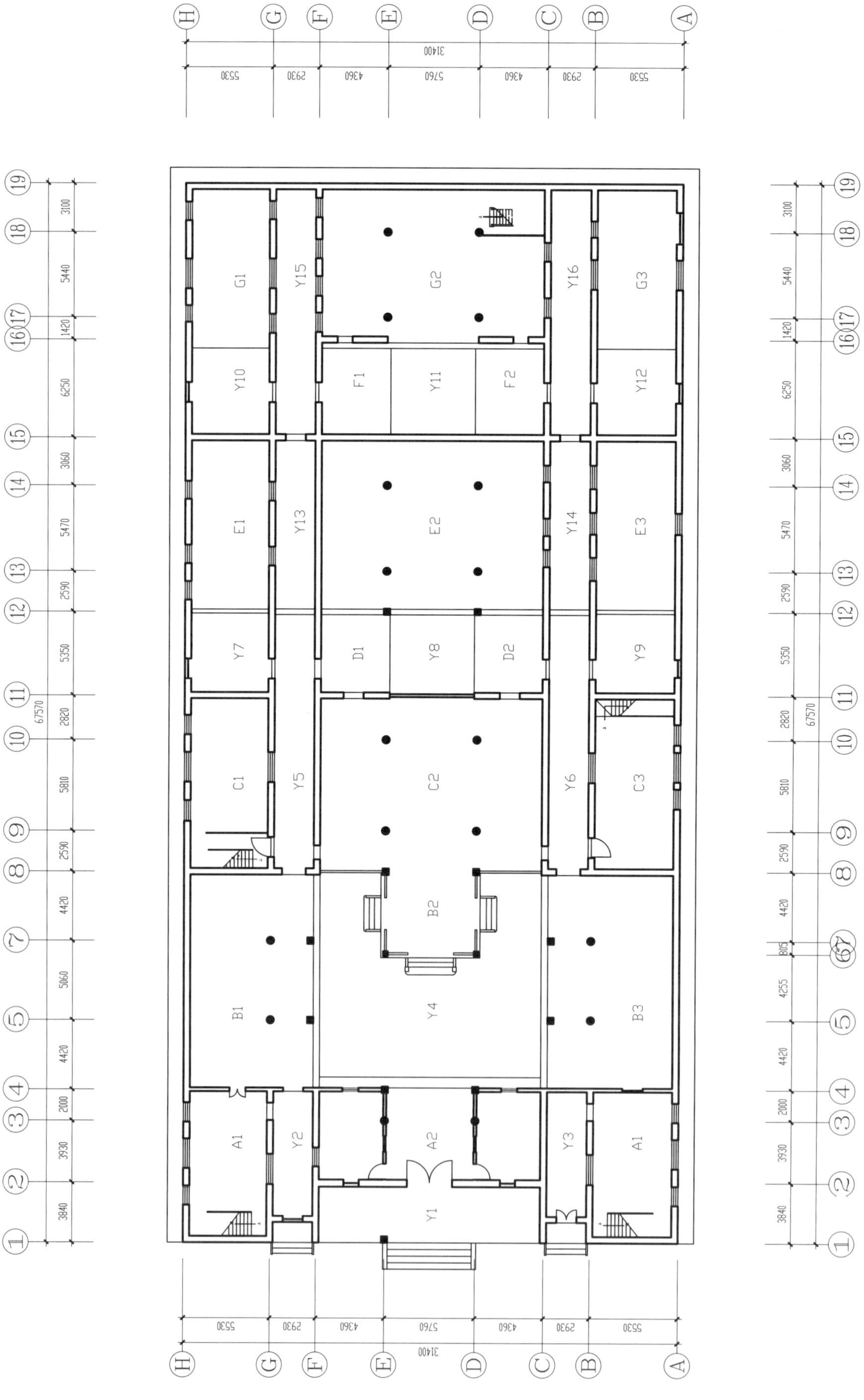
G1
Y15
G2
Y16
G3
Y10
F1
Y11
F2
Y12
E1
Y13
E2
Y14
E3
Y7
D1
Y8
D2
Y9
C1
Y5
C2
Y6
C3
B1
B2
Y4
B3
A1
Y2
A2
Y3
A1
Y1
31400
5530
2930
4360
5760
4360
2930
5530
67570
3100
5440
1420
6250
3060
5470
2590
5350
2820
5810
2590
4420
5060
4420
2000
3930
3840
805
4255

房间院落编号

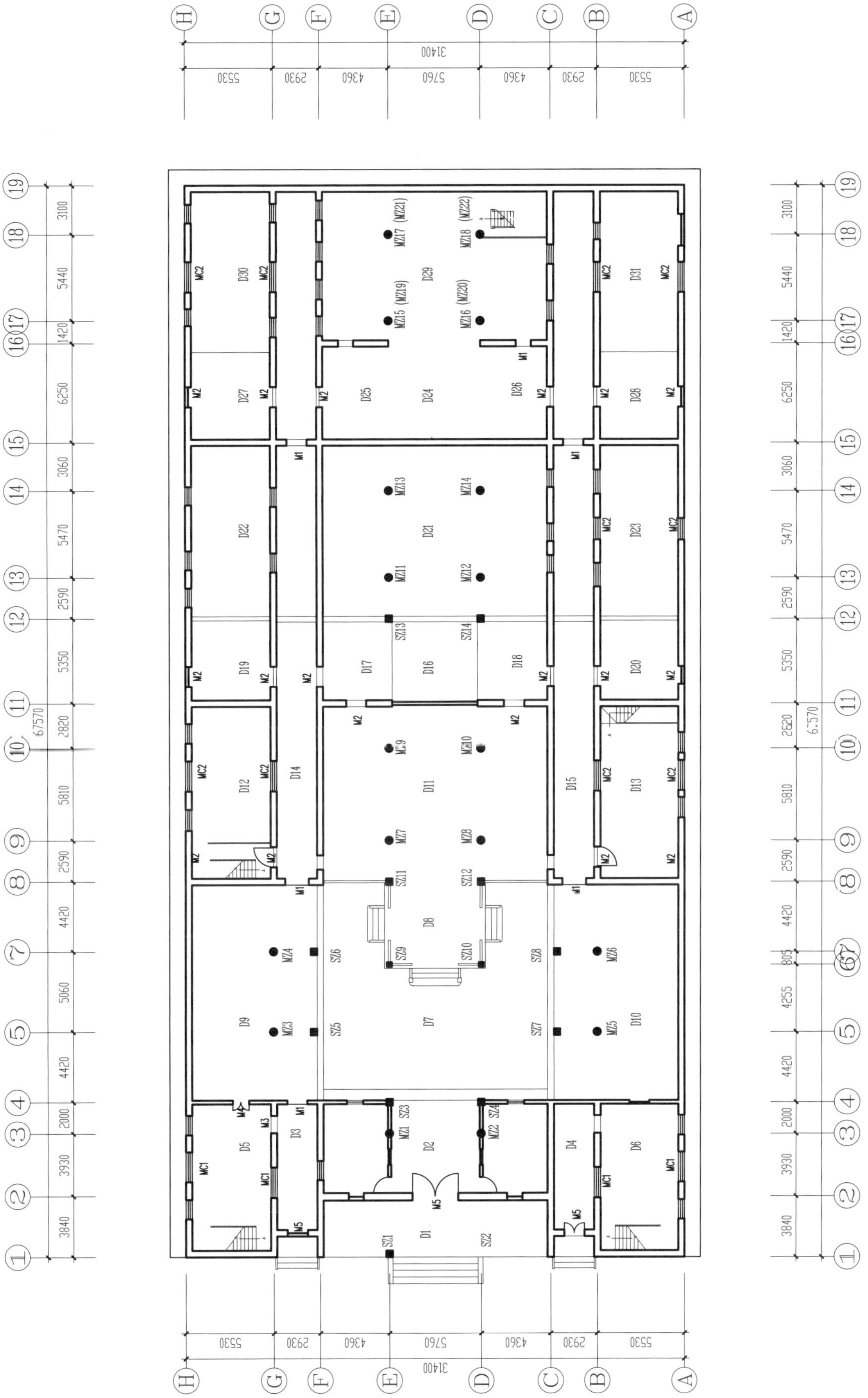
H G F E D C B A
31400
5530 2930 4360 5760 4360 2930 5530
1 2 3 4 5 6 7 8 9 10 11 12 13 14 15 16 17 18 19
3840 3930 2000 4420 5060 4420 2590 5810 2820 5350 2590 5470 3060 6250 1420 5440 3100
805 4255
67570
D1 D2 D3 D4 D5 D6 D7 D8 D9 D10 D11 D12 D13 D14 D15 D16 D17 D18 D19 D20 D21 D22 D23 D24 D25 D26 D27 D28 D29 D30 D31
SZ1 SZ2 SZ3 SZ4 SZ5 SZ6 SZ7 SZ8 SZ9 SZ10 SZ11 SZ12 SZ13 SZ14
MZ1 MZ2 MZ3 MZ4 MZ5 MZ6 MZ7 MZ8 MZ9 MZ10 MZ11 MZ12 MZ13 MZ14 MZ15 (MZ19) MZ16 (MZ20) MZ17 (MZ21) MZ18 (MZ22)
M1 M2 M3 M4 M5 MC1 MC2

地面、柱、门窗编号

修复设计图纸

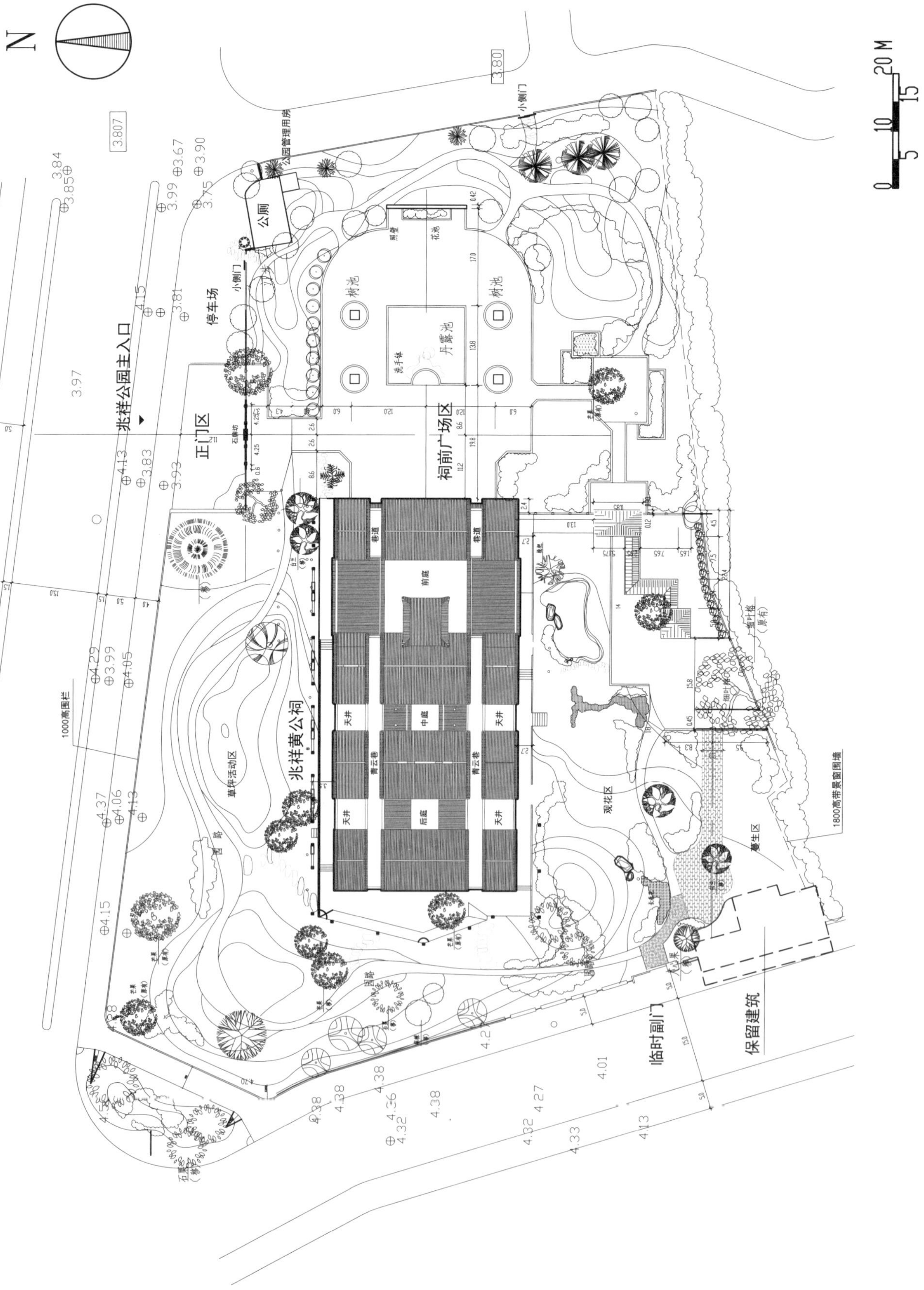

兆祥公园总平面图

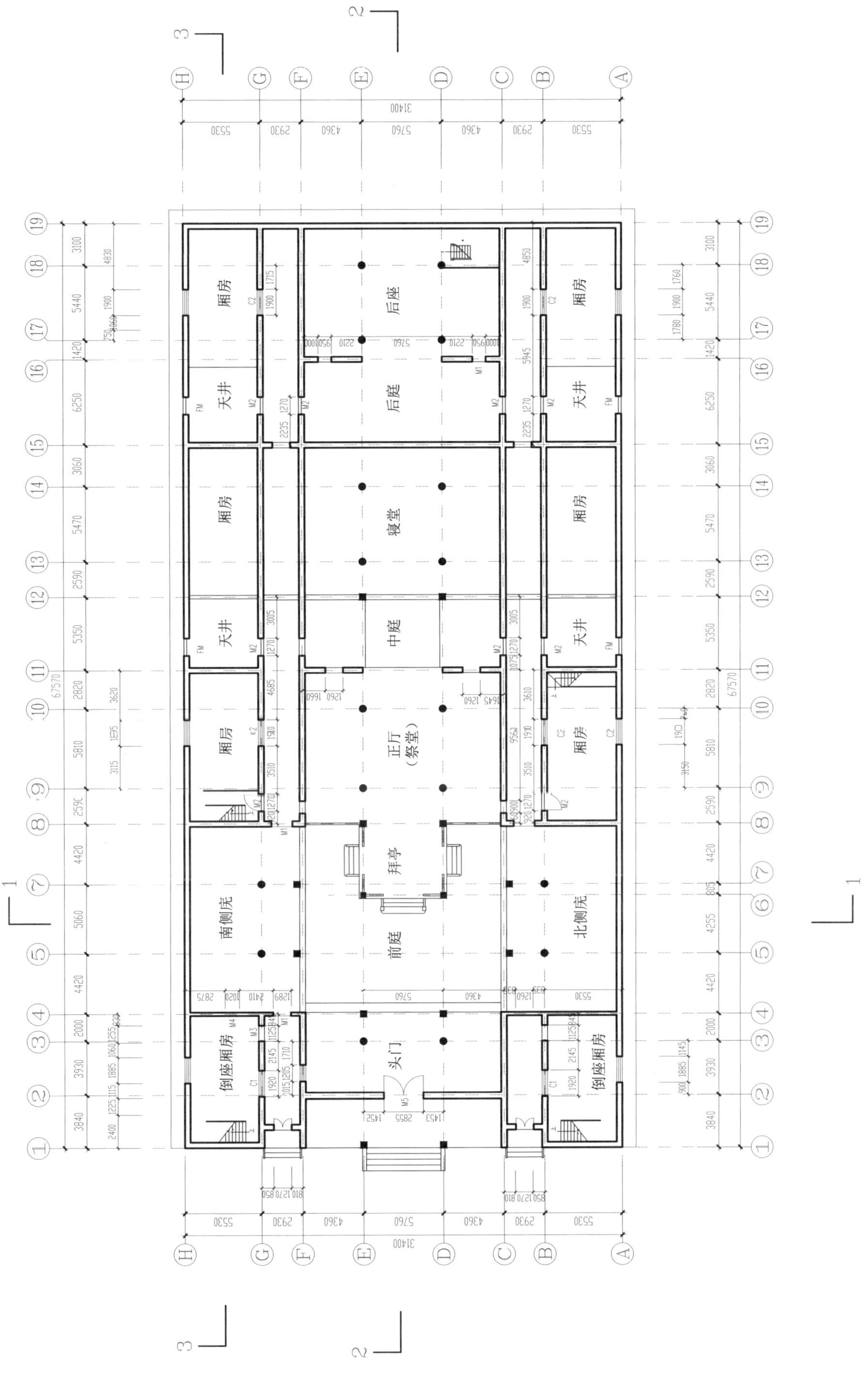

首层平面图

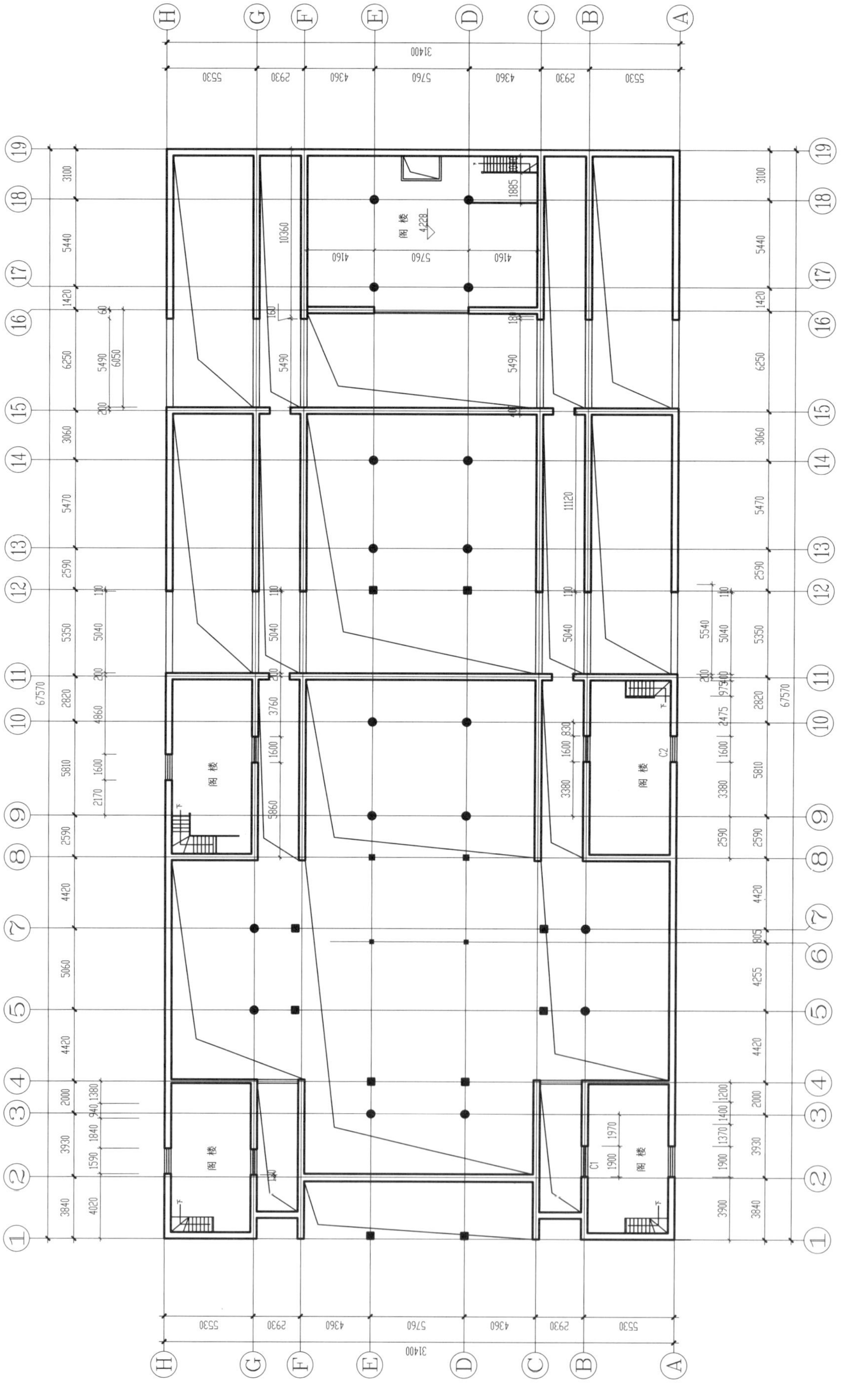

阁楼层平面图

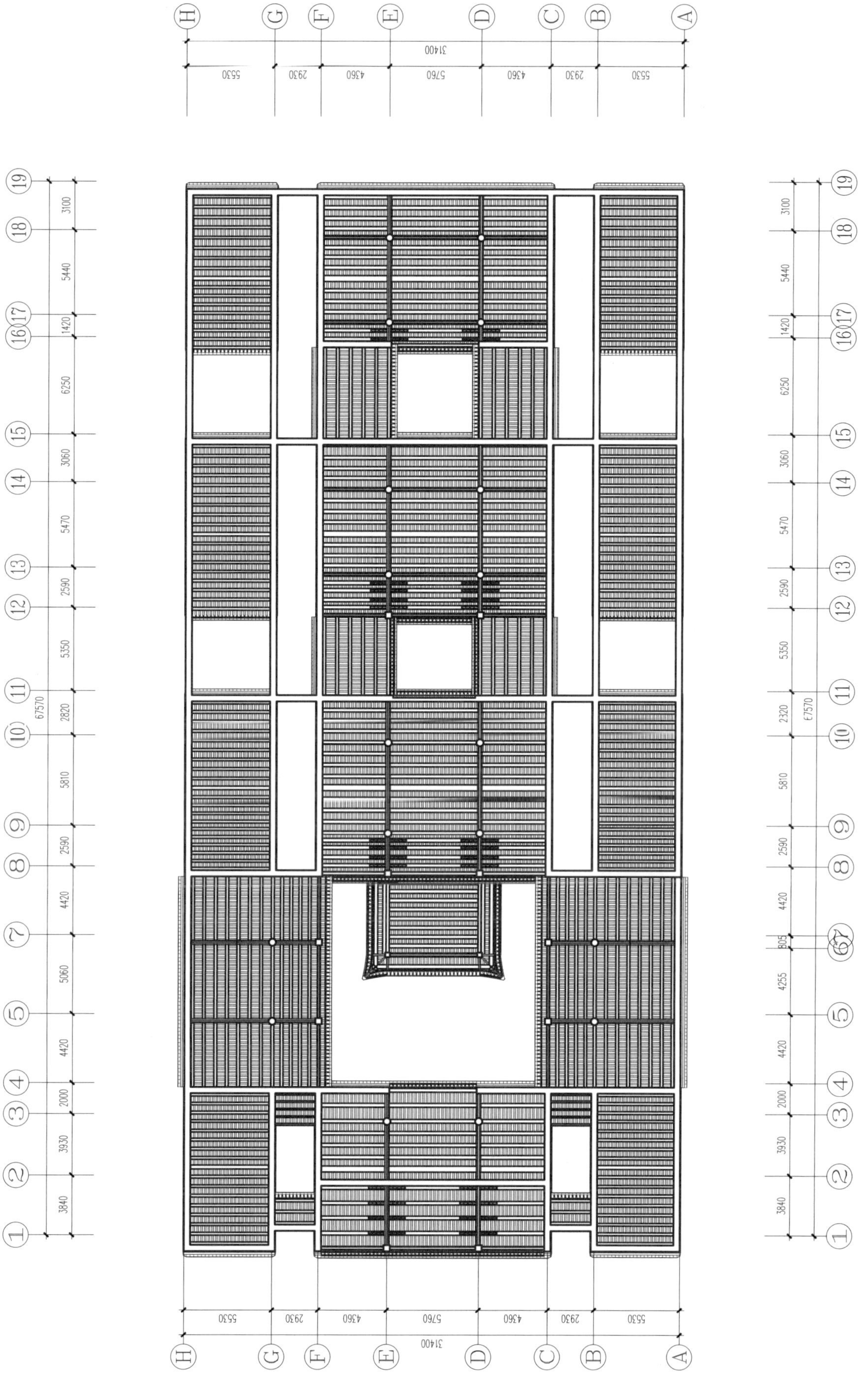
H
G
F
E
D
C
B
A
31400
5530
2930
4360
5760
4360
2930
5530
1
2
3
4
5
6
7
8
9
10
11
12
13
14
15
16
17
18
19
3840
3930
2000
4420
5060
4420
2590
5810
2820
5350
2590
5470
3060
6250
1420
5440
3100
67570
4255
805

屋顶仰视平面图

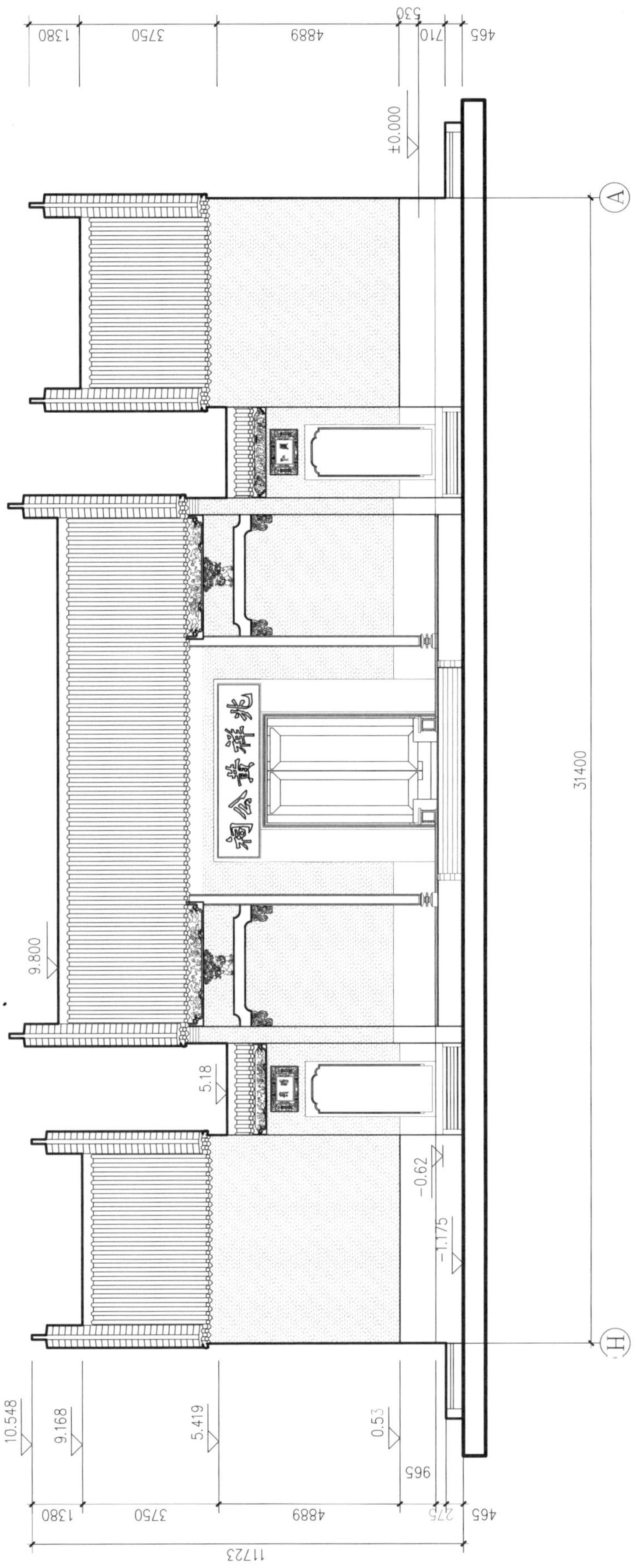
1380
3750
4889
530
710
465
±0.000
A
H
31400
兆祥黄公祠
9.800
5.18
-0.62
-1.175
10.548
9.168
5.419
0.53
965
275
11723

正立面修复设计图

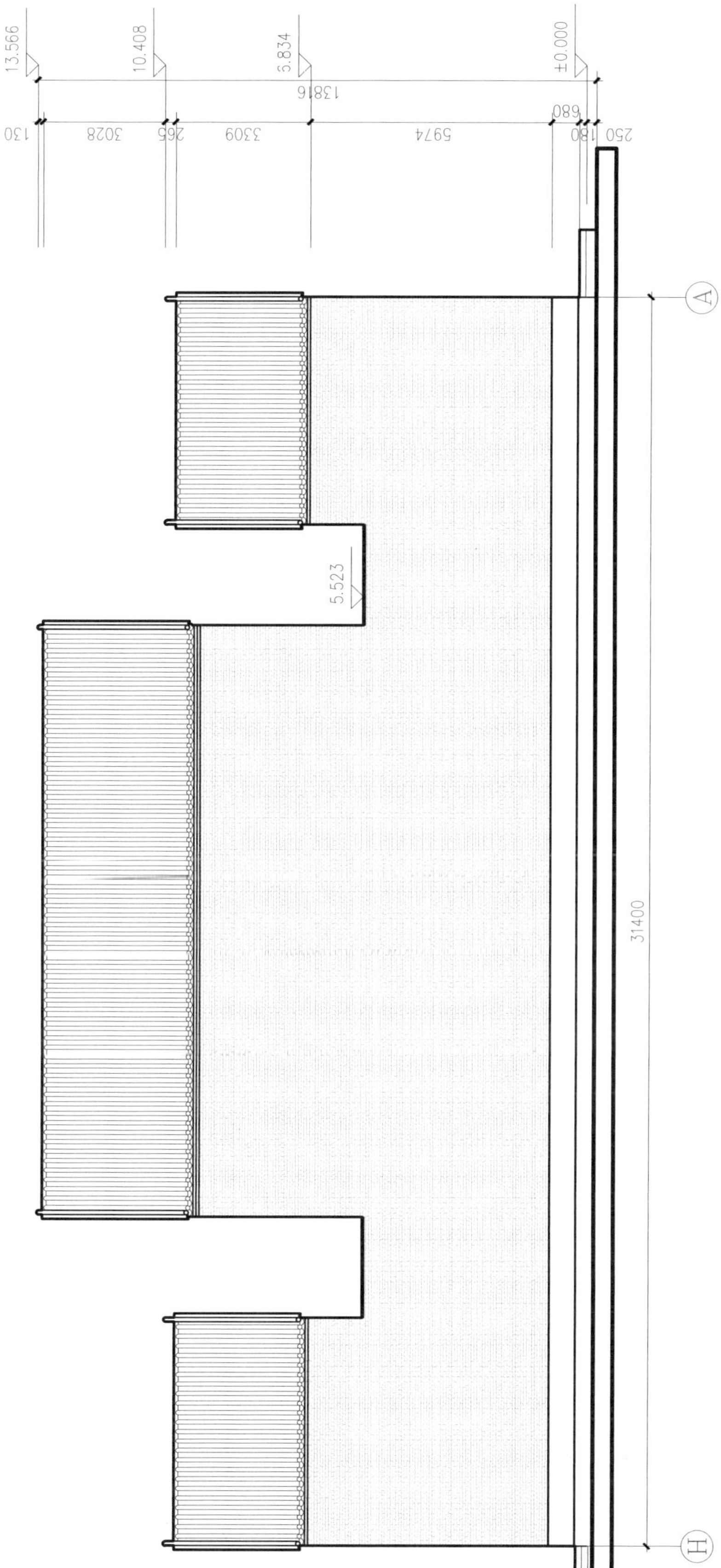
13.566
10.408
5.834
±0.000
13816
130
3028
265
3309
5974
680
180
250
5.523
31400
A
H

背立面修复设计图

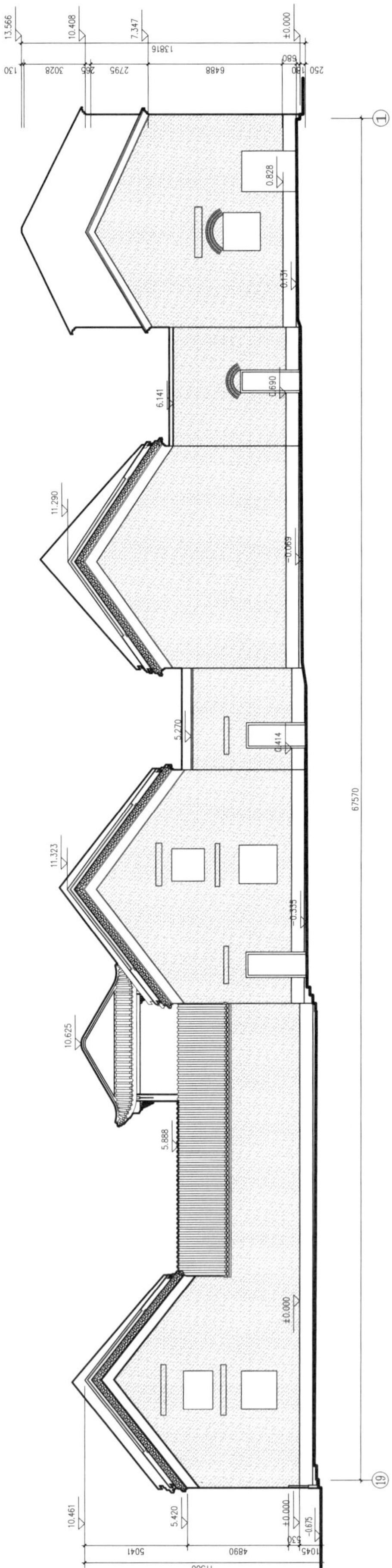
13.566
10.408
7.347
±0.000
13816
130
3028
265
2795
6488
680
250
180
①
0.828
0.131
6.141
0.690
11.290
-0.069
5.270
0.414
67570
11.323
-0.335
10.625
5.888
±0.000
⑲
10.461
5.420
±0.000
-0.675
5041
4890
530
1045
11506

北立面

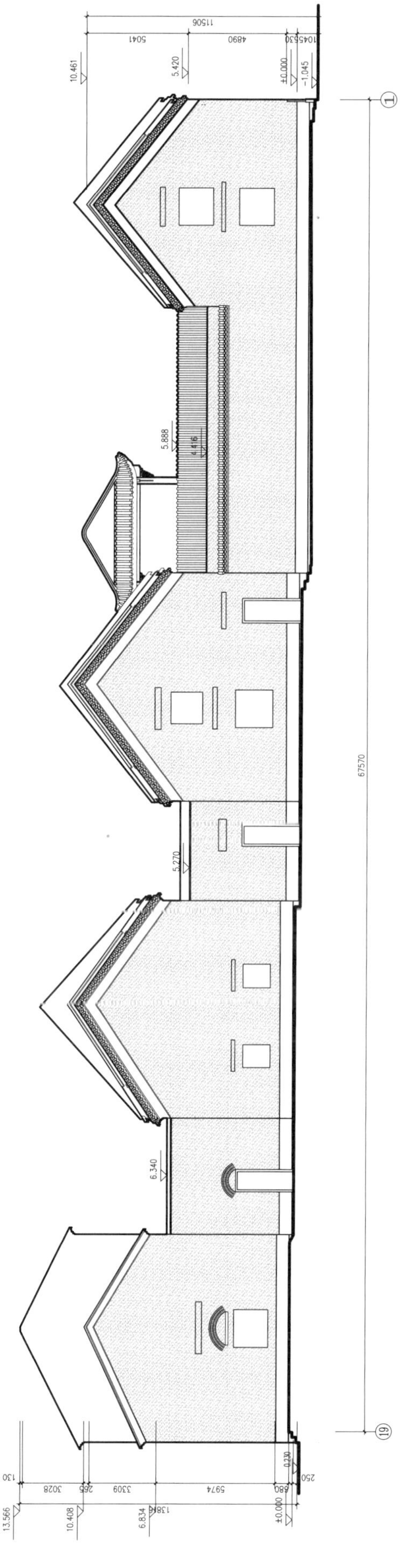
11506
5041
4890
10.461
5.420
±0.000
-1.045
67570
5.888
4.416
5.270
6.340
13.566
10.408
6.834
±0.000
130
3028
265
3309
5974
680
250
230
1384

南立面

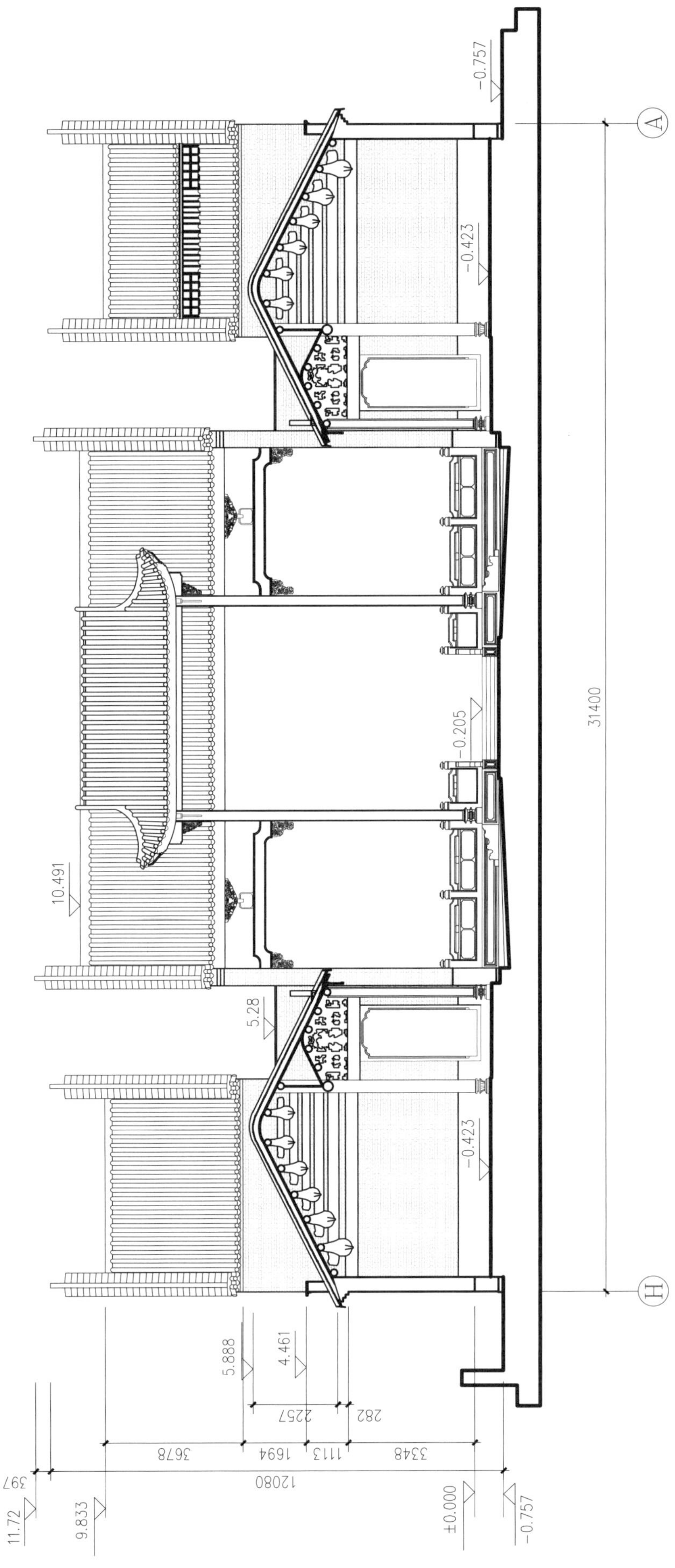
A
H
31400
-0.757
-0.423
-0.205
10.491
5.28
5.888
4.461
2257
282
3678
1694
1113
3348
12080
397
11.72
9.833
±0.000
-0.757

1-1 剖面

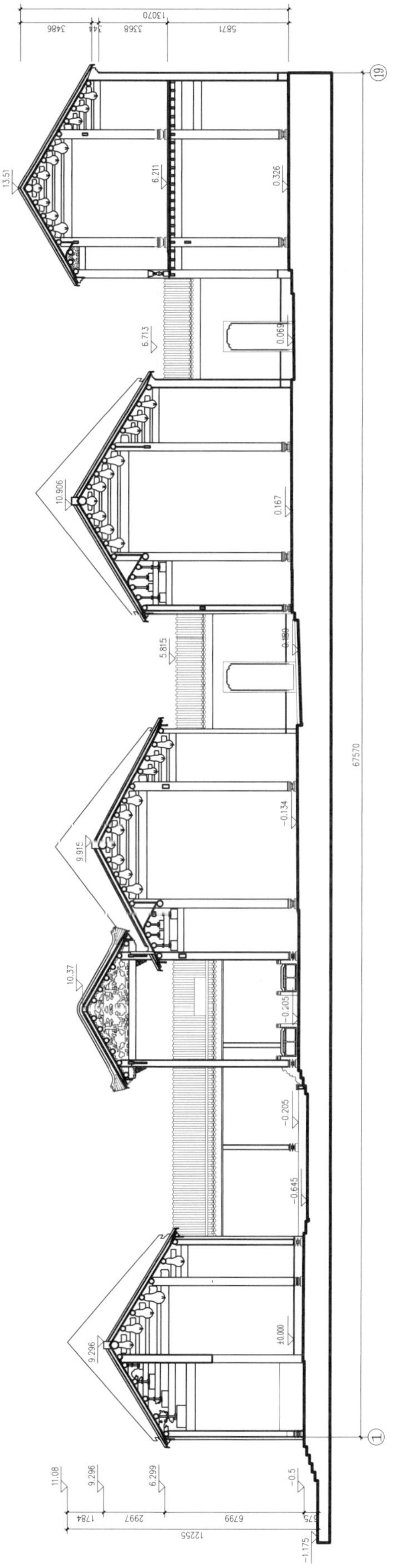

2-2 剖面

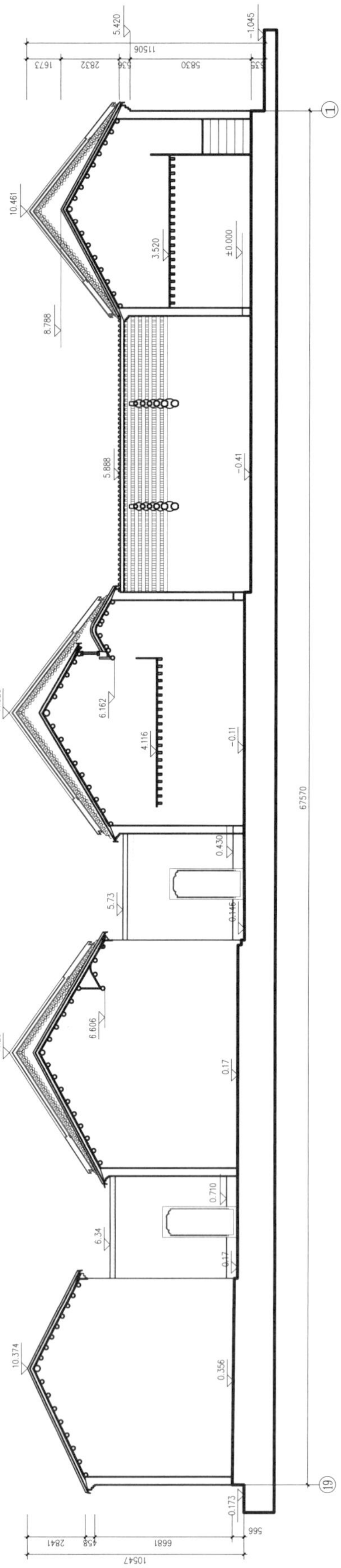
5.420
-1.045
11506
1673
2832
536
5830
535
10.461
3.520
±0.000
8.788
5.888
-0.41
11.323
6.162
4.116
-0.11
0.430
5.73
0.146
11.29
6.606
0.17
0.710
6.34
0.17
10.374
0.356
-0.173
67570
2841
458
6881
995
10547

3-3 剖面

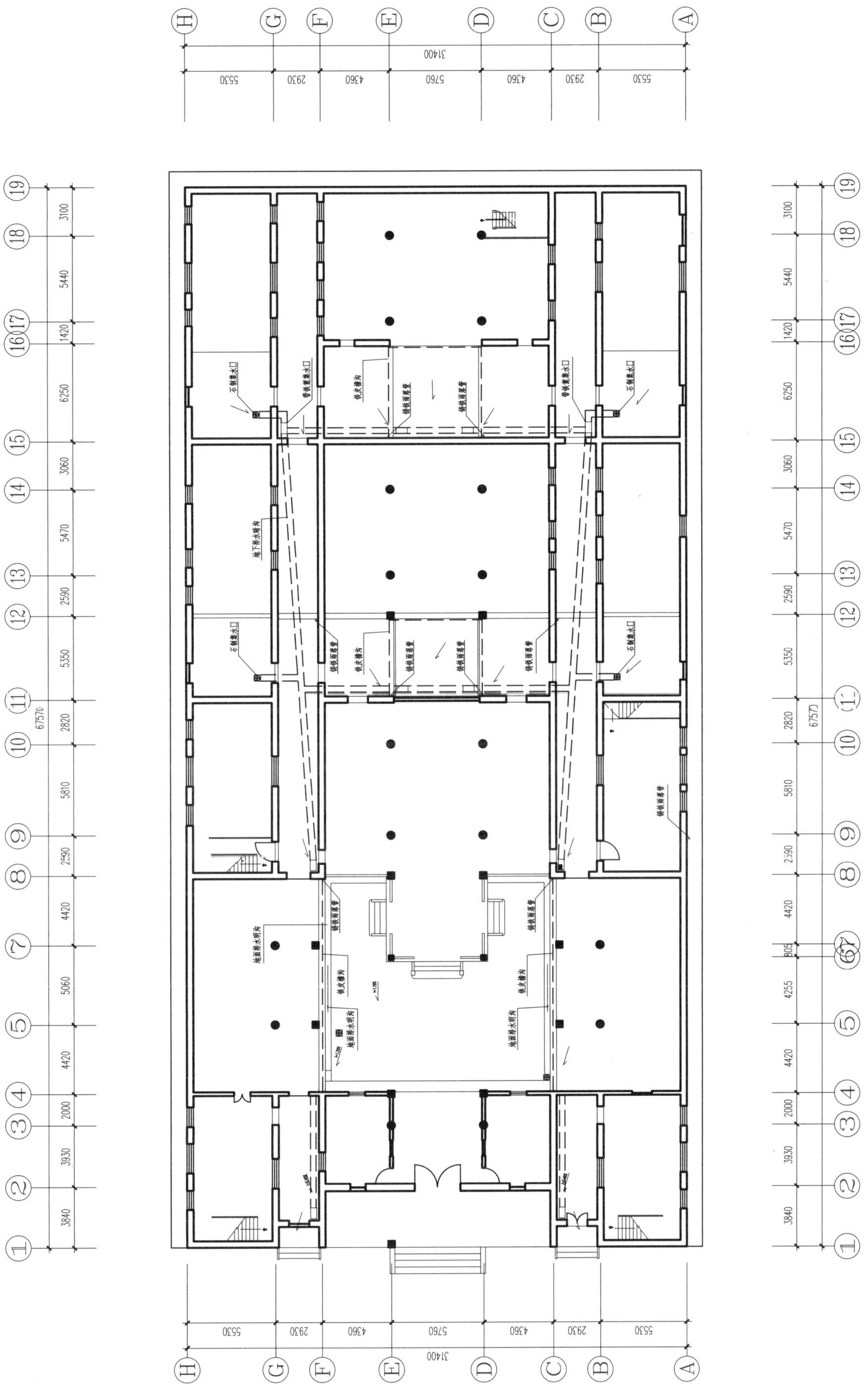

排水系统平面图

门窗大样图

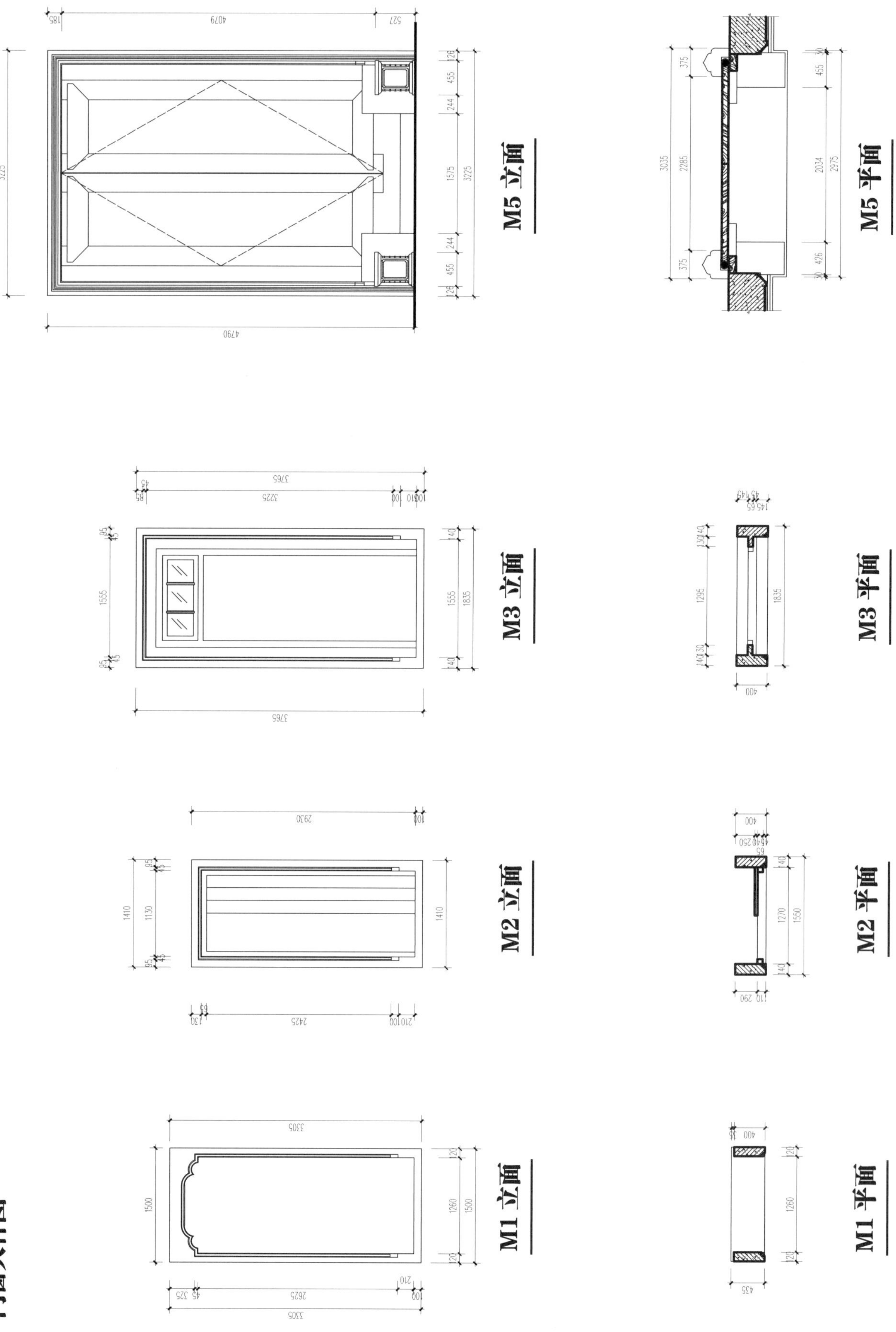

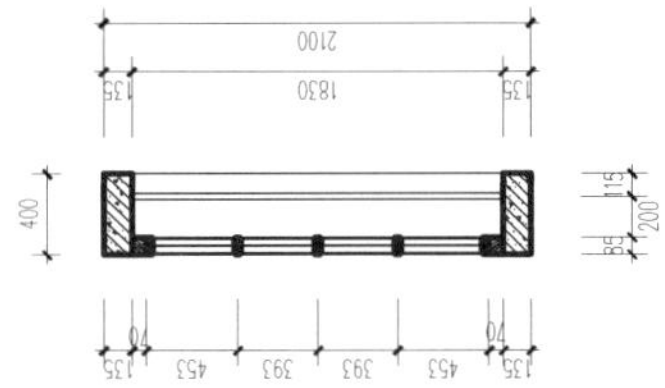
C2 剖面

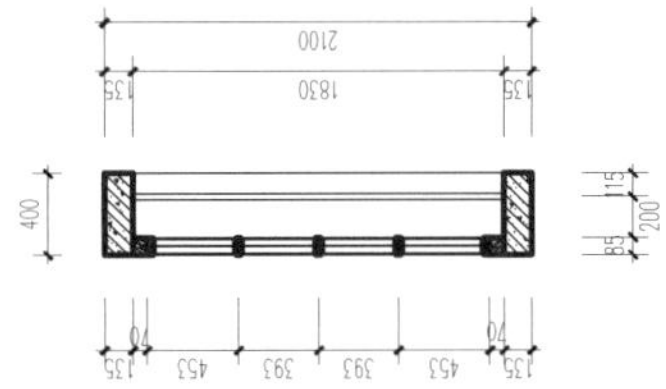
C1 剖面

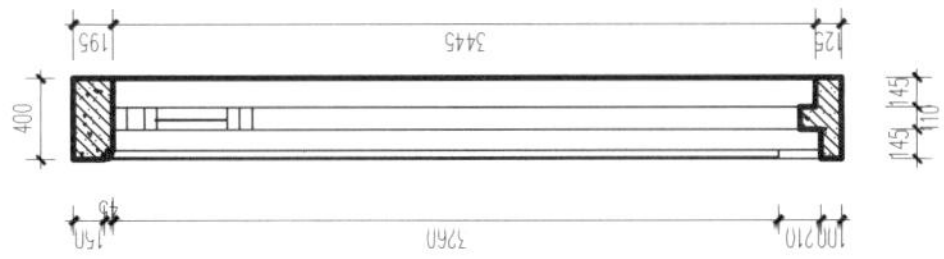
M3 剖面

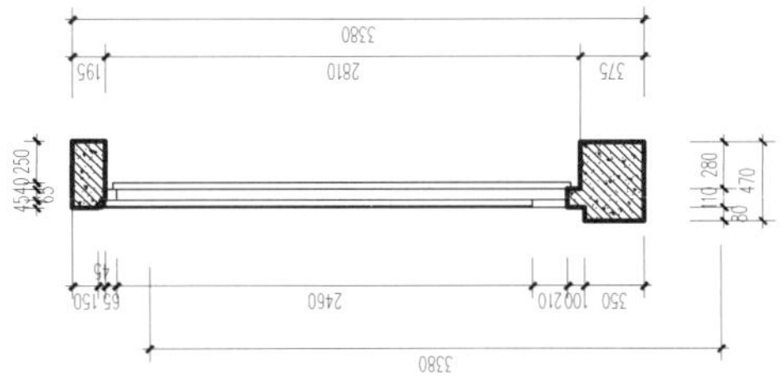
M2 剖面

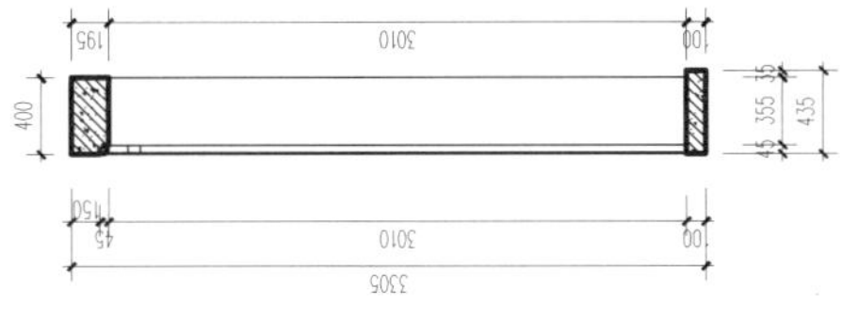
M1 剖面

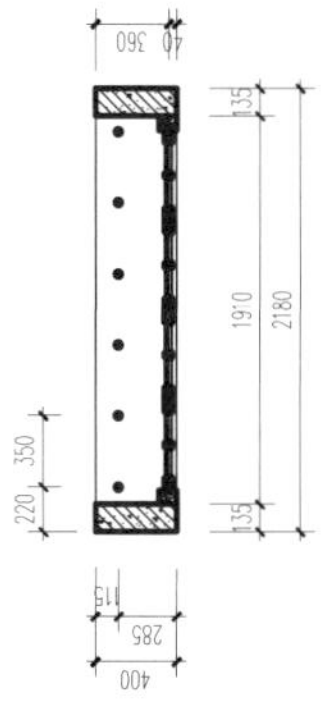
C2 平面

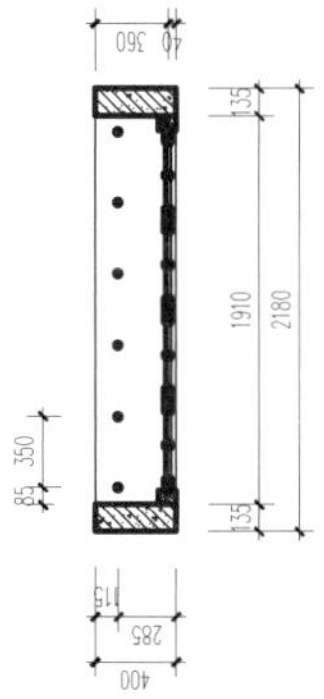
C1 平面

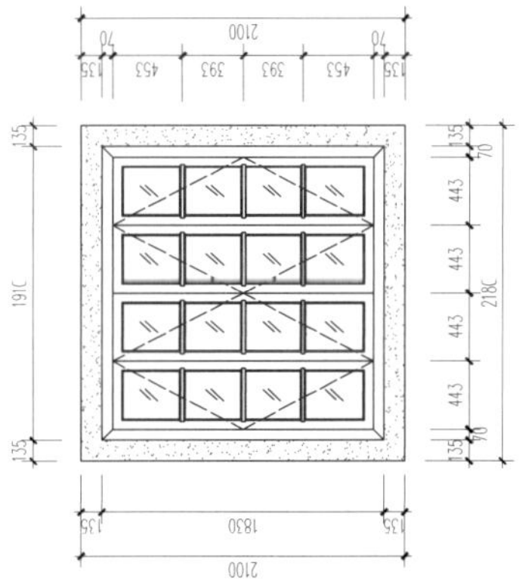
C2 立面

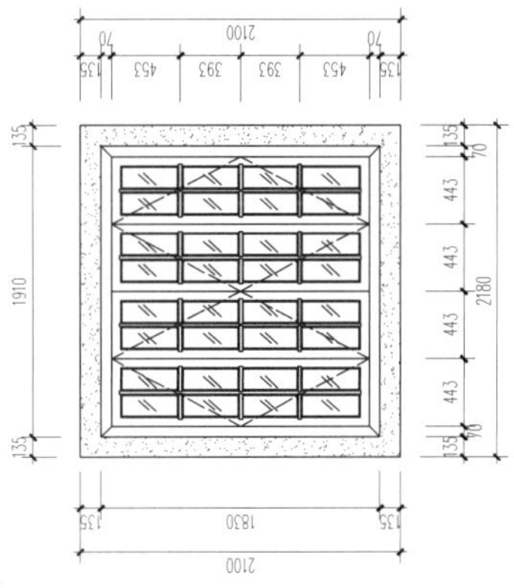
C1 立面

柱础大样图

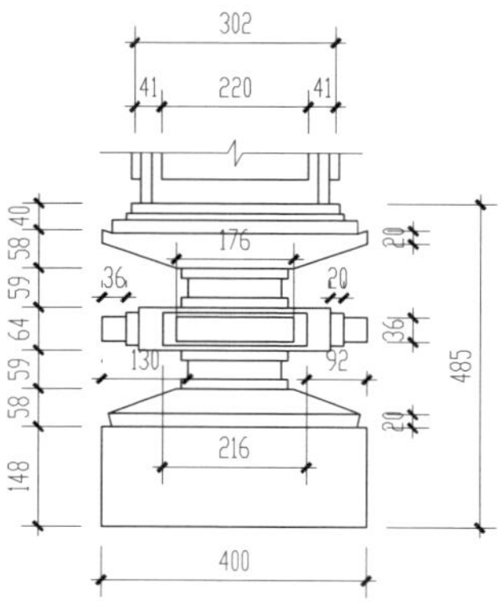

正门柱础西立面

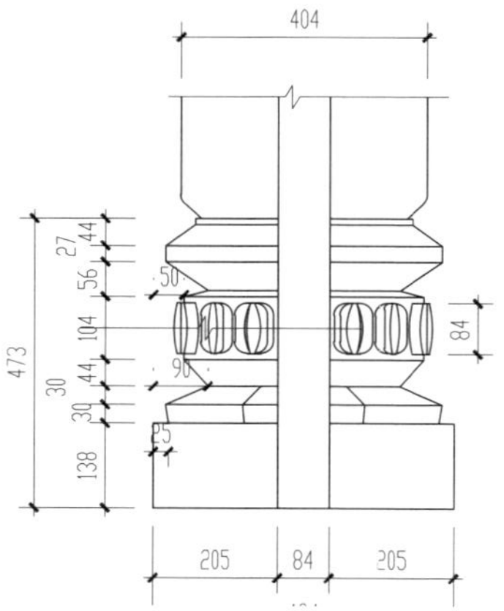

正门内中轴北侧柱立面

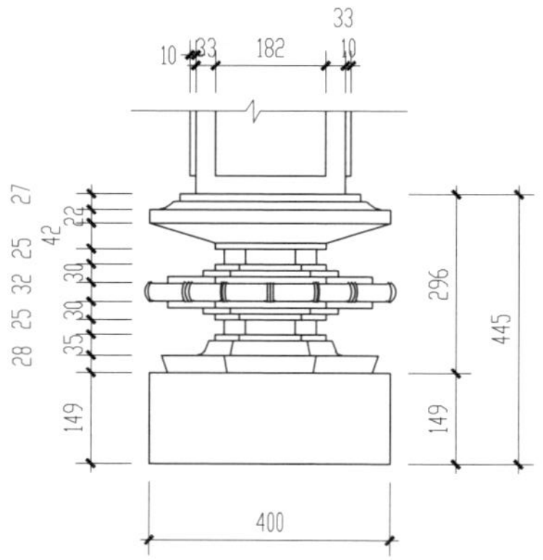

一进东侧柱础东立面

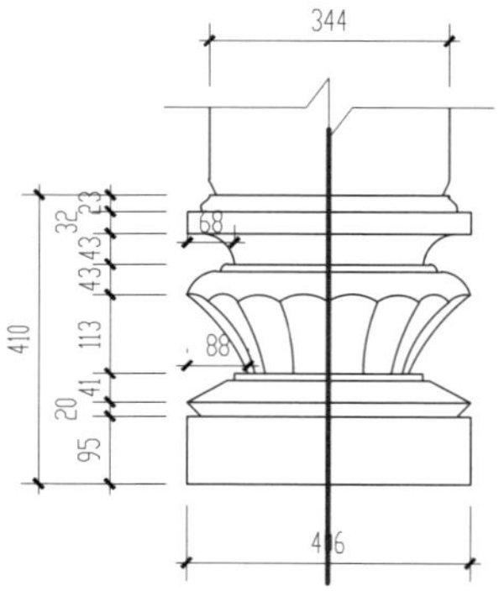

一进北侧廊内厢房柱础立面

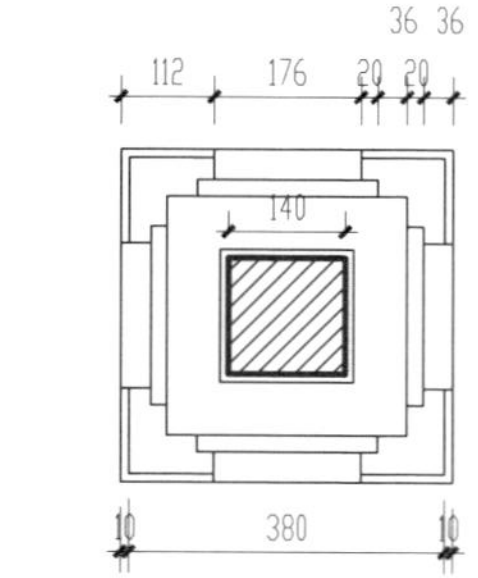

正门柱础 I-I 剖面

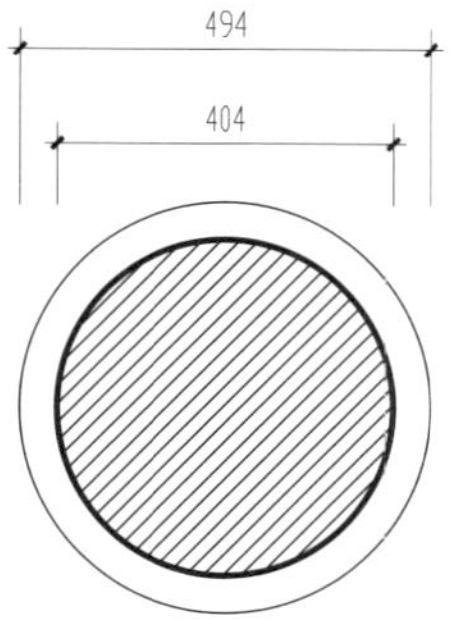

正门内中轴北侧柱剖面

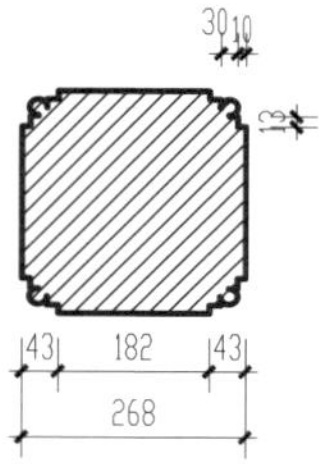

一进东侧柱剖面

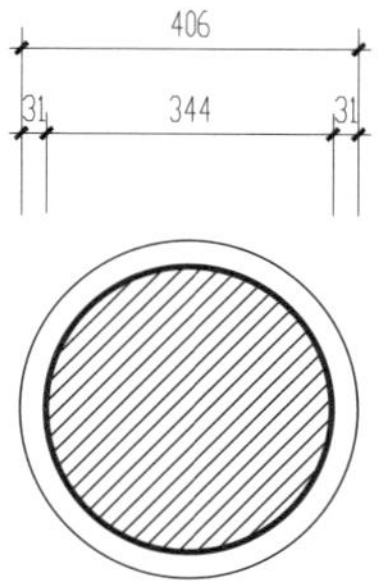

一进北侧廊内厢房柱础剖面

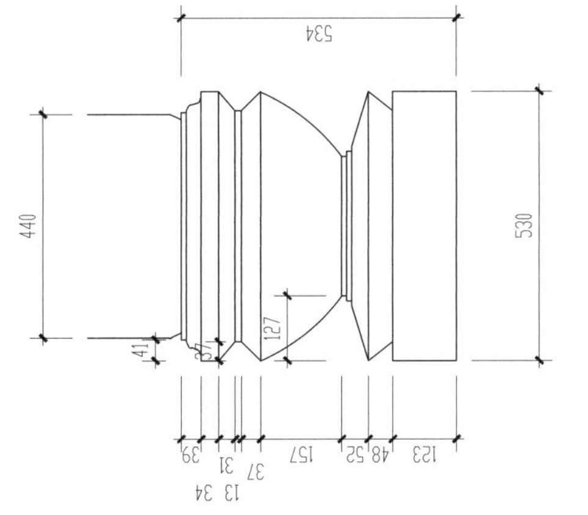

四进二层柱础立面

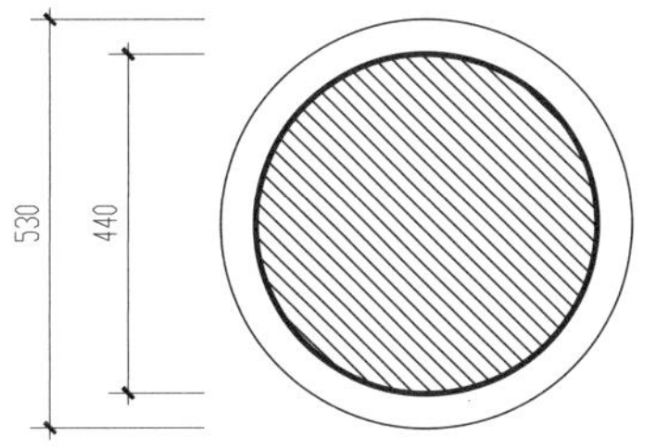

四进二层柱础剖面

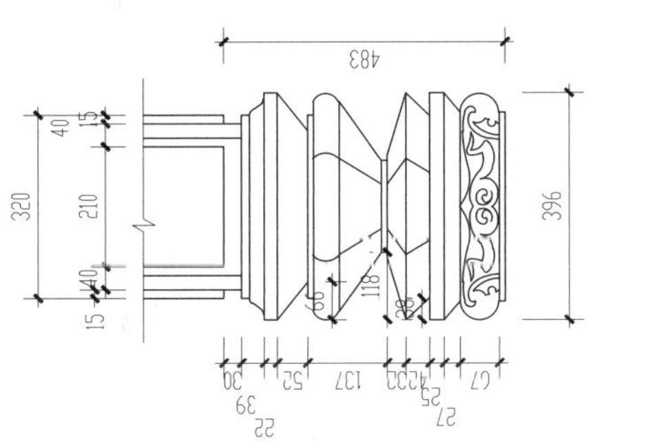

三进内柱础立面

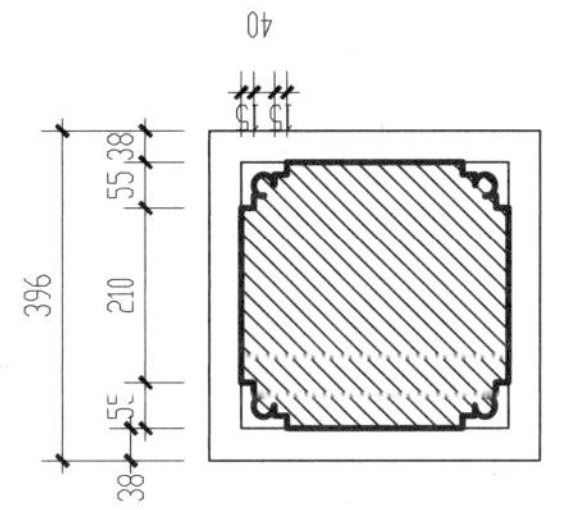

三进内柱剖面

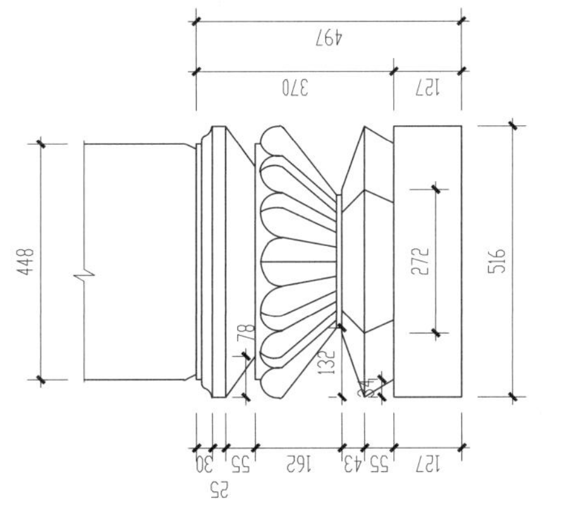

二进内柱础立面

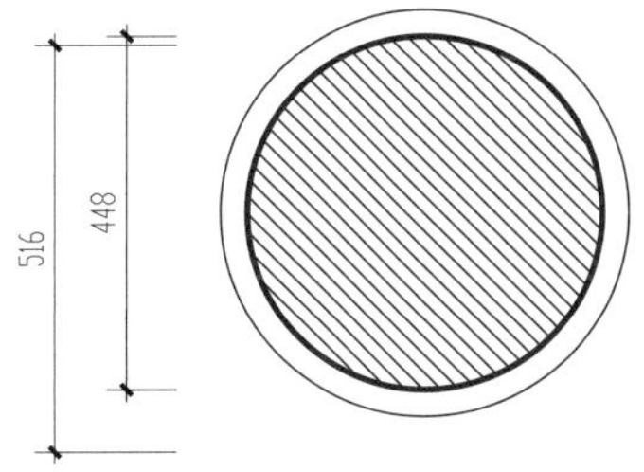

二进内柱剖面

5103
680
1248
1248
1248
680
56
174
307
143
143
307
165
18
165
R291
R300
16
16
100
100
56
78
32
266
198
38
199
267
32
114
56
668
668
1334

三进厅内挂落

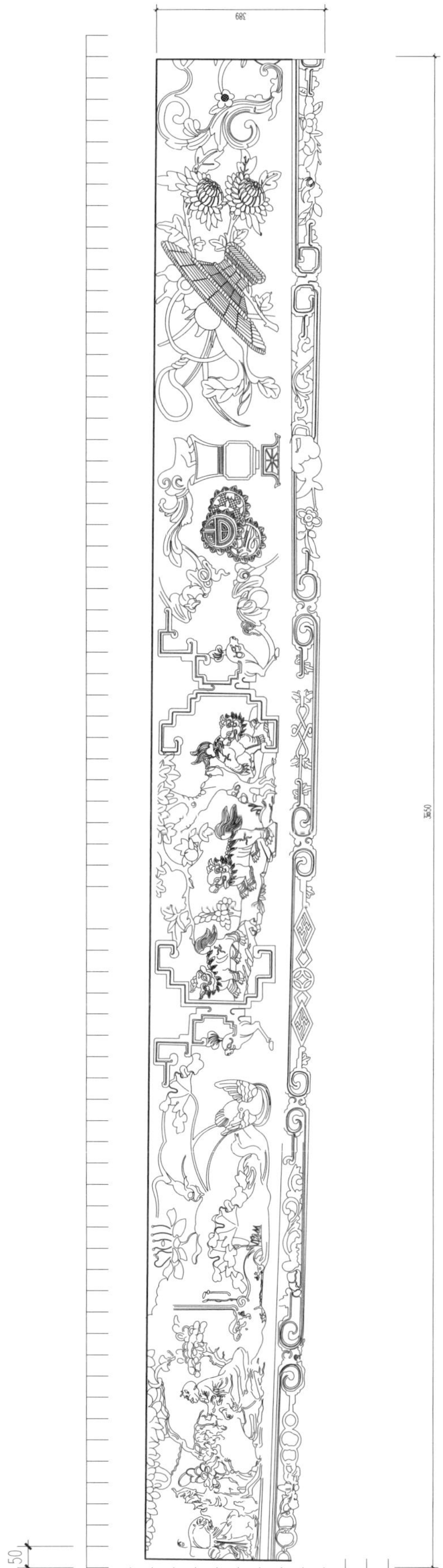

封檐板大样（局部）

三、申报文件与获奖情况

UNESCO
Asia-Pacific
Heritage
AWARDS
for
Culture Heritage Conservation

an entry for

UNESCO ASIA-PACIFIC HERITAGE AWARDS 2004
for Culture Heritage Conservation

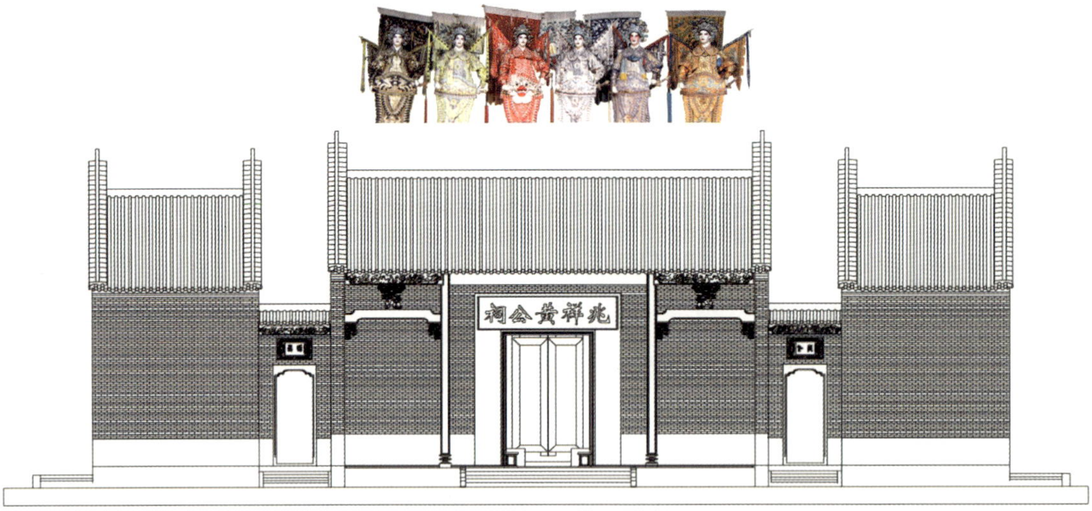

CONSERVATION OF ZHAOXIANG HUANG ANCESTRAL HALL

Foshan, Guangdong, China

UNESCO
Culture in Asia and the Pacific

an entry for

UNESCO ASIA-PACIFIC HERITAGE AWARDS 2004
for Culture Heritage Conservation

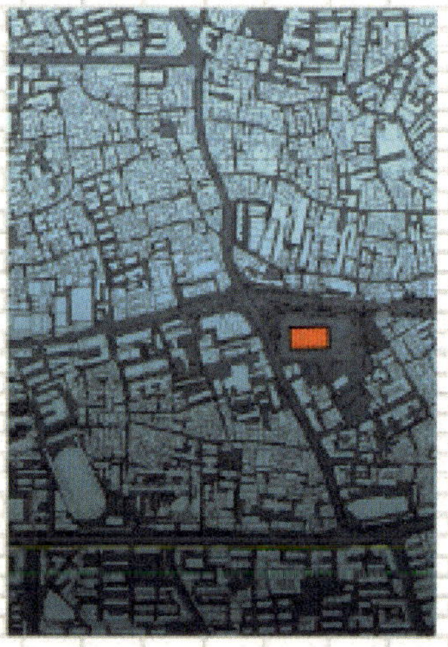

CONSERVATION OF ZHAOXIANG HUANG ANCESTRAL HALL

Foshan, Guangdong, China

CONSERVATION OF ZHAOXIANG HUANG ANCESTRAL HALL, **Foshan, Guangdong, China**

UNESCO ASIA-PACIFIC HERITAGE AWARDS
for Culture Heritage Conservation
2005 ENTRY FORM

To: Office of the Regional Advisor for Culture in Asia and the Pacific
United Nations Educational Scientific and Cultural Organization
UNESCO Bangkok Office
920 Sukhumvit Road, Bangkok 10110, THAILAND
Tel: (66 2) 391 0577, 391 0550 ext 509
Fax: (66 2) 391 0866
E-mail: culture@unescobkk.org

Date of Submission: 25, Apr, 2005

PLEASE TYPE:

Name of Project *(Entry)*

Conservation of Zhaoxiang Huang Ancestral Hall

Project is submitted for consideration to: (Tick ONLY one box)

■ 2005 UNESCO Heritage Awards for Culture Heritage Conservation
□ 2005 UNESCO Heritage Awards Jury Prize for Innovation

Location of Project *(Street, City, Postcode, State, Country)*
Zhaoxiang Park, Zhaoxiang Road, Foshan, Guangdong, China

Date of Completion of the Project *(must be after October 1994)* 29 Jan, 2002

Name or Names of Individuals Responsible for the Project

Professor Wu Qingzhou

Name of Building Owner whose permission for Nomination has been obtained *(if not submitted by owner)*

Urban Construction Bureau, Foshan, Guangdong (Property trustee)

Address of Owner

528000, No.98, Renmin Road, Chancheng District, Foshan, Guangdong

Name(s) of Architect/Designer/Consultant

Pro. Wu Qingzhou, Feng Shu, Feng Jiang, Ou Jie, Zhou Yigang, Xiao Min

Address(es) of Architect/Designer/Consultant

510640, Academy of Oriental Archi-culture, South China University of Technology, Guangzhou

CONSERVATION OF ZHAOXIANG HUANG ANCESTRAL HALL, **Foshan, Guangdong, China**

Name of General Contractor

the First Building Construction Company, Wuhua, Guangdong

Address of General Contractor

501507, Room 401, No.26, Yandu Road, Tianhe District, Guangzhou, China

SUBMITTED BY:

Name Wu Qingzhou

Address 510640, Academy of Oriental Archi-culture, South China University of Technology, Guangzhou

City Guangzhou Postal Code 510640 Country China

Phone Number (0086-20-87597501) Fax Number (0086-20-87114884)

Mobile Phone Number (0086-0-13609794568) Email jarher@163.com

How did you learn about the Asia-Pacific Heritage Awards?

From Internet and Magazine,

CHECKLIST:

The following documents are enclosed:

a. ■ Drawing, including site plan, floor plans and sections. (New, old, and reconstructed parts should be clearly marked in colour on the drawings). Total no. of enclosed:__________

b. ■ 5" x 7" Photographs used to support the project description, clearly labelled with location and identified on a plan. Total no. enclosed: [min. 15]__________

c. ■ Project Description (in English), 6 to 8 (A4 sheets):

d. ■ Project Completion Report *(if available)*

e. ■ Owner Consent

f. ■ Occupant Comments

g. ■ Additional documentation *(not required)*

FRONT VIEW :   BEFORE and AFTER

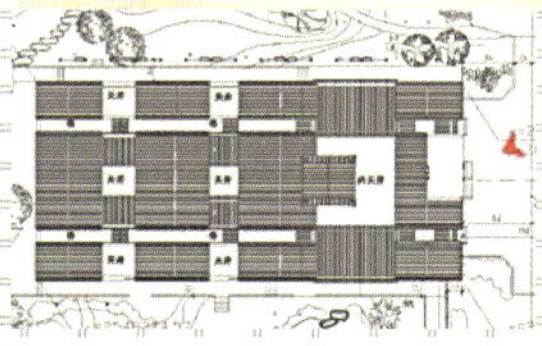

FRONT VIEW OF THE 1st SECTION : BEFORE and AFTER

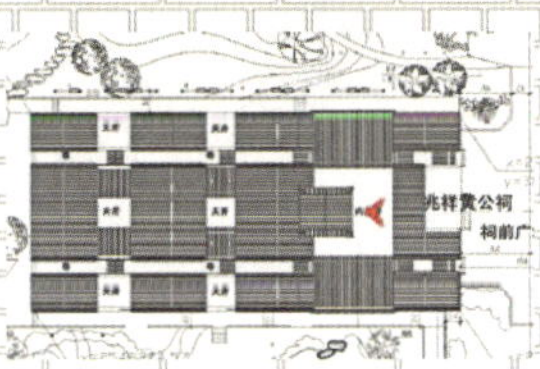

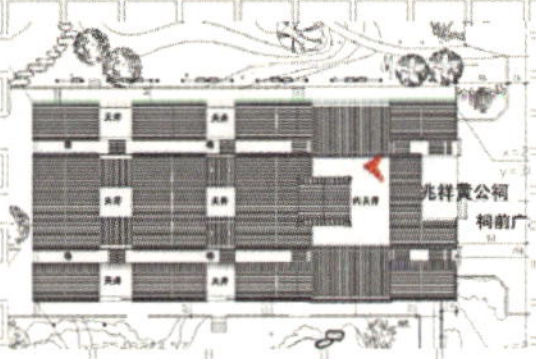

TIMBER-FRAME OF THE 3rd SECTION: BEFORE and AFTER

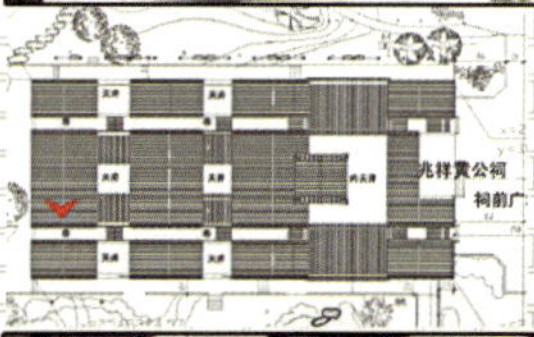

GABLE WALL :   BEFORE and AFTER

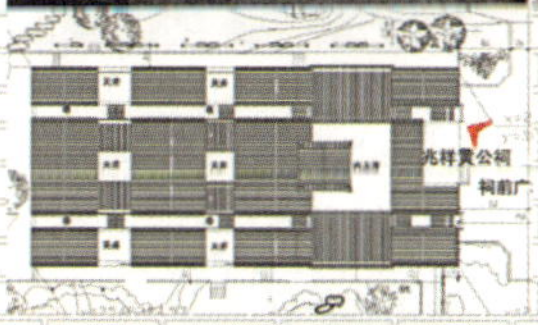

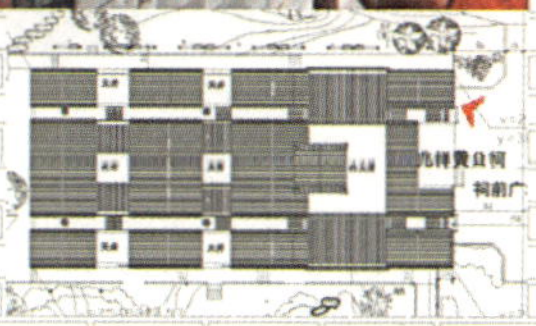

STONE-FRAME OF THE MAIN GATE: BEFORE and AFTER

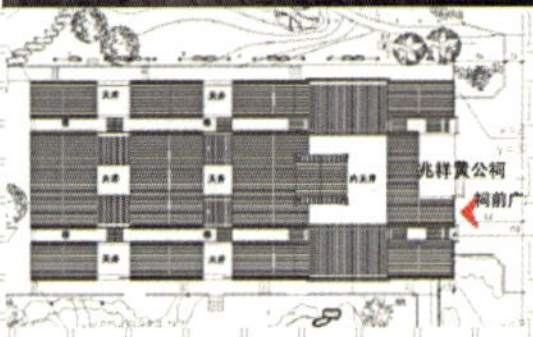

BACK SIDE OF THE MAIN GATE: BEFORE and AFTER

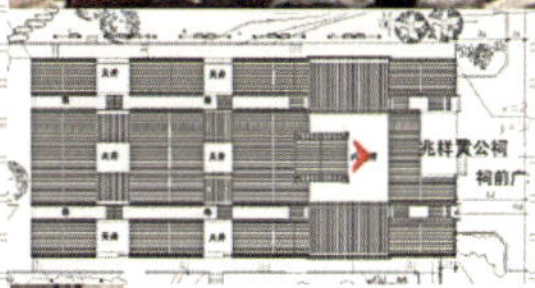

TIMBER-FRAME OF THE 1st SECTION: BEFORE and AFTER

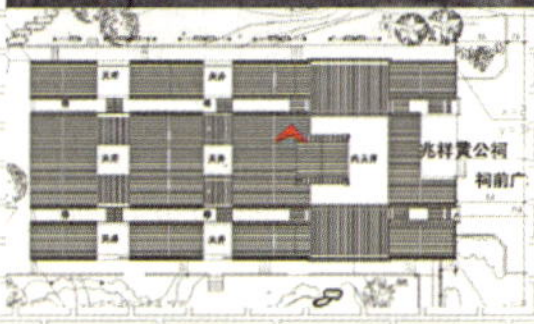

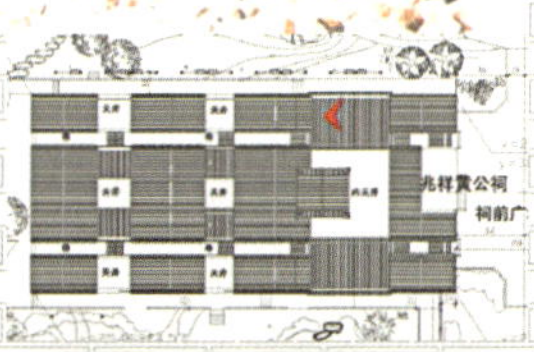

GABLE WALL OF THE 1st SECTION : BEFORE and AFTER

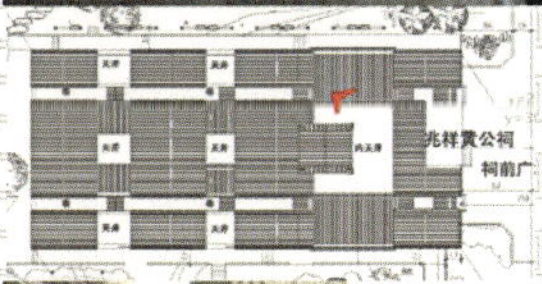

GABLE WALL OF THE 1st SECTION : BEFORE and AFTER

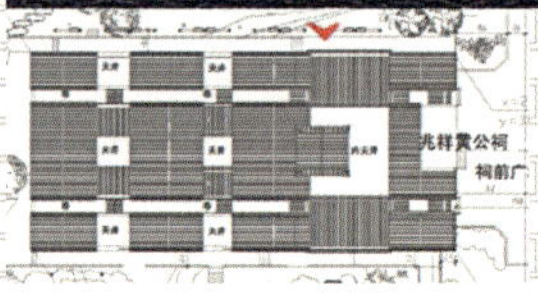

WOOD CARVING OF THE MAIN GATE : BEFORE and AFTER

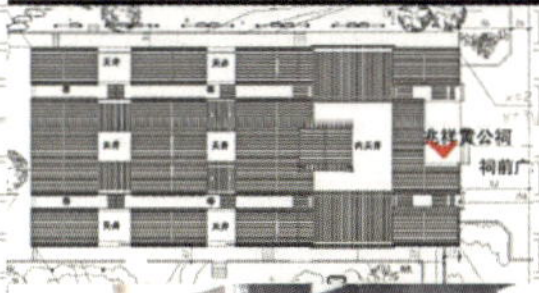

CONSERVATION OF ZHAOXIANG HUANG ANCESTRAL HALL, Foshan, Guangdong, China

□ **Synopsis of conservation project**

Standing at the traditional city center of Foshan, Zhaoxiang Huang Ancestral Hall is a shrine in memory of Huang Danian, the initiator of renowned Chinese traditional medicine "Huang Xianghua Medicated Balm". It is the largest ancestral hall nowadays in the downtown of Foshan. The Hall falls on the historic development axis of Foshan City, and locates southeast of the crossroad where Funing Rd meets East Jianxin Rd. The construction of the Hall lasted 15 years, from 1905 (the 31st year of Guangxu Reign, Qing Dynasty) to 1920 (the 9th year of the Republic of China). Positioned in the west and facing the east, the Hall covers a land area of about 3,060m^2, with the architecture 67.57meters in length and 31.40meters in width. The programming features courtyards of 3 sections, where the main architectures like the Main Gate, the Alter Pavilion and the Main Hall are all placed along the central axis, on both sides flank symmetrically the wing-rooms which are connected to the main architectures through laneways, showing the typical ancestral hall typology in the Pearl River Delta. The whole architectural complex, with its magnificent verve, elaborate design and rich ornaments, is the most representative among the ancestral halls in Foshan with higher historic and artistic values.

During the decades of historical vicissitude, the Hall could still be fortunate enough to survive wars, social-cultural changes and the massive urban construction and was basically kept intact. Now it is in the list of Provincial Historic Reservation Units. Sponsored jointly by the non-governmental organizations and the governmental sectors, the restoration of the Hall was officially started up in 2000. The conservation project adopted a multilateral participation mode, and established a multilateral conference mechanism involving the property right trustee, future user, research organizations, warm-hearted cultural personnel, etc. The surveying, research, restoration design, construction and supervision were all implemented in strict accordance with the technical procedures. With the associated efforts of various parties involved, the restoration project lasting 2 years was successfully completed in 2002, and the Hall was handed over to Cantonese Opera Museum in 2003 for services, which attracted close attention and concern from the all walks of lives and the wide range coverage of the news media. Now It is among the very few well-operated museums of local opera in China.

Pay respect to
the founder of Cantonese Opera

Local opera show
in the inauguration of Cantoness Opera Museum

Location of Project in China

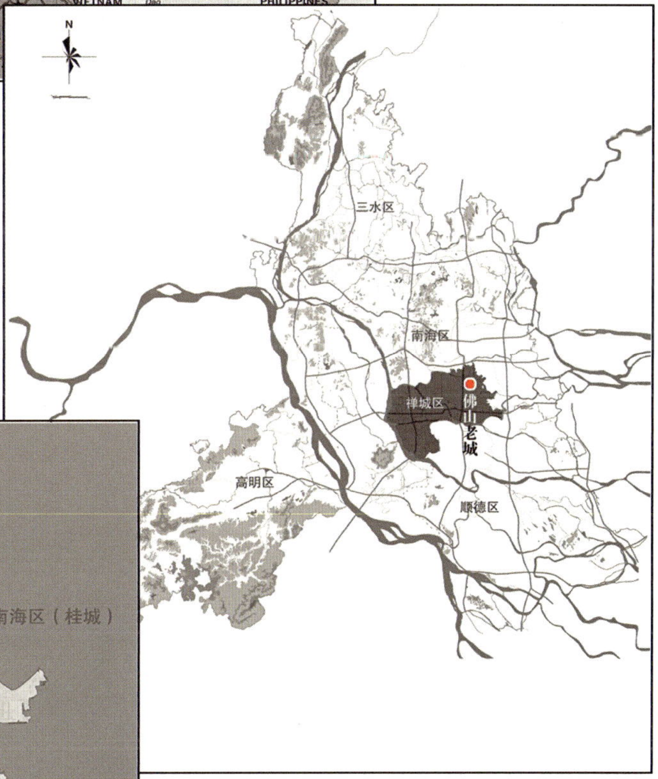

Location of Project in City

Location of Project in Traditional City Center

□ Conservation project's goals and objectives (client's brief)

The objective of this project aims to, on the precondition of ensuring the structural safety of the Hall and by displaying the original and genuine characters of the architecture, to preserve the valuable cultural heritages and historical information, therefore to make the Hall a memorial place of Huang Family, and even the whole traditional Chinese Medicine industry of Foshan. Meanwhile, the program also aims to provide a key social space of great significance to the city for showcasing Foshan's traditional culture, continuing the function as a dynamic public space and cultural carrier in the city and community of today.

The project aims to explore how to use the limited resources, how to differentiate this restoration program from the improper way of using coined ornaments in other ancient architecture protection cases in this area, and how to take correct measures to provide more effective protection and maintenance to the cultural relic that had experienced numerous additions and reconstructions, hence re-emphasize the values of the traditional craftsmanship and the traditional architecture, accumulate practical experiences for the regional cultural relics protection, and eventually make spatial and cultural contributions to the public life.

Foshan is the seminary of Cantonese Opera . The renovated Hall is used as "Cantonese Opera Museum". It's now the only Cantonese Opera Museum in China. All funding and exhibits are raised and collected from the non-governmental channels, including donation from home and abroad. The Museum, aiming to display the origin, development and worldwide spreading of the Cantonese Opera, is open to the public. From June 2003 to December 2004, the Museum had received visitors totaling more than 150,000 person/times.

Daily show by fans of local opera

CONSERVATION OF ZHAOXIANG HUANG ANCESTRAL HALL, Foshan, Guangdong, China

The Hall in the Traditional City Centre (Now)

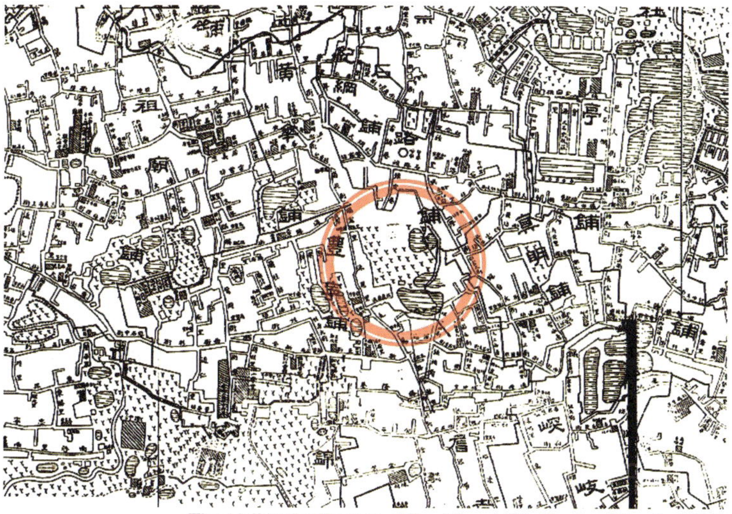

The Hall in the Texture of Traditional City Centre (Map in 1923)

□ The background of the property

- **Historical background**

As one of the cradles of Traditional Chinese Medicine , Foshan boasts tremendous correlative historical relics in the traditional downtown, one of which is Zhaoxiang Huang Ancestral Hall.The Hall had been once used as the office of Kuomintang Party Committee of Nanhai County, afterwards was stationed in 1949 by the People's Liberation Army. During the Cultural Revolution many artistic members of the Hall were substantially destroyed. In 1985, a car driven by a soldier bumped into and broke the stone column at the Main Gate, and led to extensive collapse of the rooftop at the Main Gate. In 1992, the Hall was required to demolish by the army as dangerous building. Thanks to the associated efforts of the various parties, it was finally resolved that the army should move out while the Hall be protected in situ.

- **Present context, condition, use of the property**

The Hall now rises within the Zhaoxiang Park which is open to the general public for free. On its east lies a small plaza and a square pool, while the park's veranda and generous green space on the other sides. The Hall is currently used in good conditions.

The opening ceremony of Cantoness Opera Museum

Appreciating the decorative lantern during the Opera Cultural Festival

- **Heritage/protection status of the property**

Owning to the impacts of the various social changes and the misuses, the architectural complex was destroyed to different extents in many places.

- *Building Appearance*

The architectural complex was kept relatively complete, yet part rooftop of the Main Gate had collapsed. Meanwhile, owning to the substantial additions and alterations during the occupancy of the Army, there were added doors, windows, partition walls and roofs in many places. In addition, the later added plastering (lime mortar) was of poor and rough craftsmanship; the rectangular stone blocks pavement outdoor were broken or sunken in many spots; the ornaments of main ridges on all architectures were missing, whereas the ridge ornaments of the side walls, except on one ridge, were all destroyed and plastered with lime mortar.

- *Interior*

Obvious upheaval had been observed with the tiled flooring of the Altar Pavilion. All original furnishings such as the steles, couplets and screens were missing, left only with some pothooks. No furniture or memorial tablets were inside the buildings.

- *Structural Conditions*

Except for the Main Gate, the overall structure was not seriously damaged. However, some parts of the beams and columns suffered from serious termite attacks, hence appeared larger crevices and need reinforcement. The rotten and decayed angles of the eaves were also generally observed.

- *Technical and Mechanical System*

Water supply - a small well at west end of the southern corridor, which was dried up then;

Water drain – the original stormwater drain system was kept comparatively complete. In the alleyways there were many remains of the cast iron stormwater pipes and the iron pothooks.

The traditional city centre in old Lunar new year's painting

An opera show in Foshan

CONSERVATION OF ZHAOXIANG HUANG ANCESTRAL HALL, Foshan, Guangdong, China

□ Scope of the conservation project

- **Stages in restoration and time frame**

In Feb 2000, the conservation program was initiated and the multilateral conference mechanism involving the construction administration authority, cultural relic authority, research organizations and design institutes was established.

In Mar 2000, Prof. WU Qingzhou ,was entrusted to chair the whole restoration and design works. He set up the basic principle for the project that "all restoration measures shall be based on the studies on the architectural archeology and history, as well as the preservation of the original and genuine features in history". An expert from Guangzhou Fine Arts Institute was invited to join the team.

In May 2000, Academy of Oriental Archiculture of South China University of Technology completed the survey and mapping.

In June 2000, researches on history, culture and architecture were carried out, as well as the diagnosis of the architecture safety issues.

In May 2000, the restoration project was started.

In June 2000, discussions were hold based on the surveying, mapping and researches regarding the specific protection and restoration principles and key technical issues.

In June 2000, the protection proposal was submitted, and was subsequently approved.

In January 2001, the restoration project was completed.

In September 2004, the urban purple line was finalized to define the boundary lines of Foshan Traditional Downtown Historic Core Protection Zone in which the Hall locates.

- **Project size (in square meters) and cost (in US$)**

Land Area: 3,060m^2
Floor Area: 2,263m^2
Project budget: US$678,160.

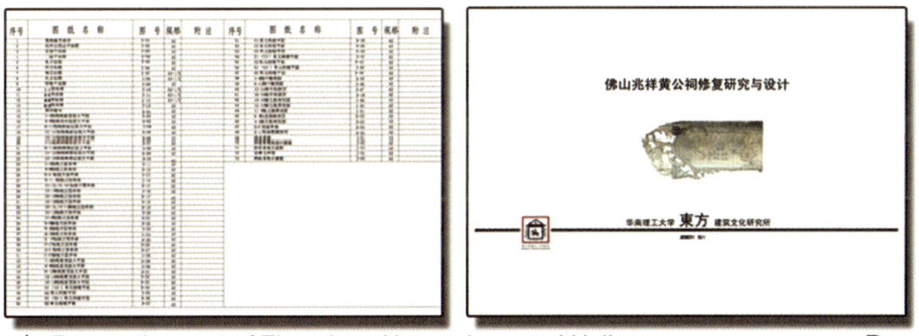
佛山兆祥黄公祠修复研究与设计

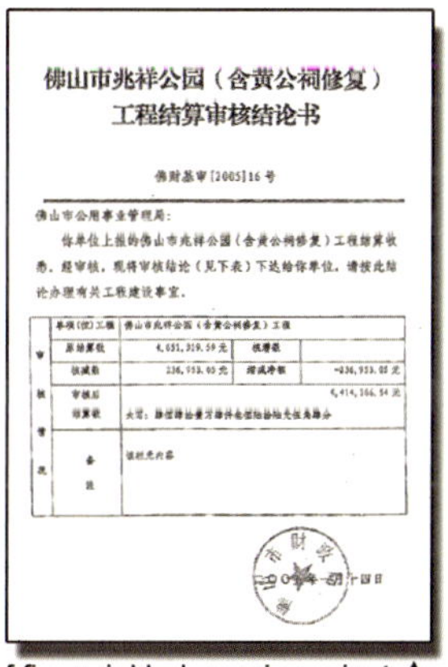
佛山市兆祥公园（含黄公祠修复）
工程结算审核结论书

▲ Research report of Zhaoxiang Huang Ancestral Hall

Report of financial balance in project ▲

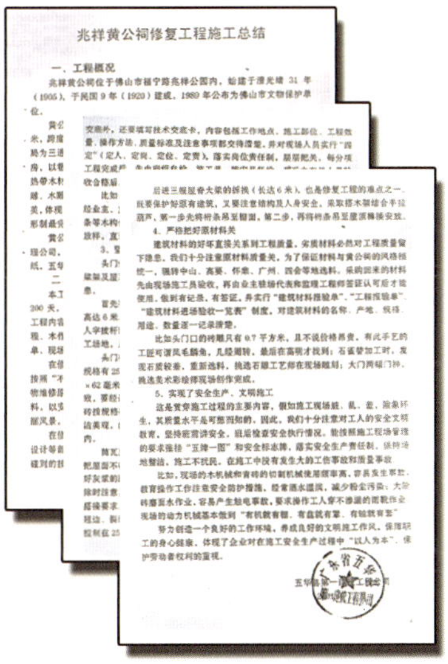
兆祥黄公祠修复工程施工总结

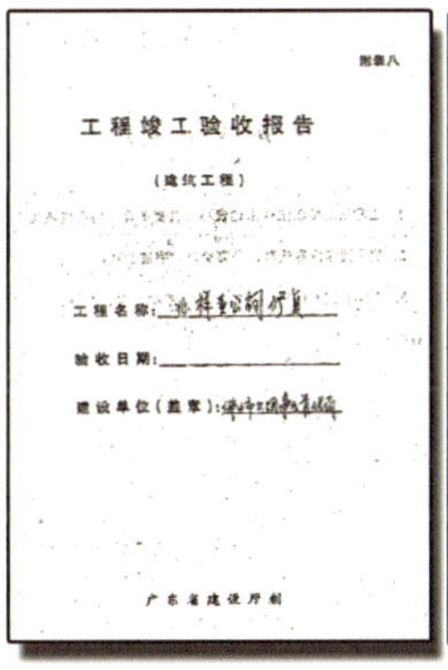
工程竣工验收报告
（建筑工程）

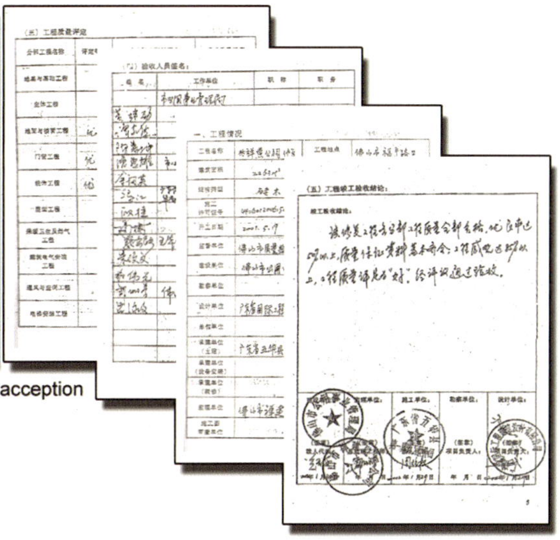

▲ Report of checking and acception

◀ Summarize of construction

□ Articulation of heritage values and significance

- Built in 1920, the Hall has a history of 85 years.
- The Hall was originally owned by Huang Family, while is now managed by the cultural affairs authority under entrustment and rented by non-governmental organizations.
- The protection works was completed in 2001, which complies with the requirement to complete the project within the recent 10 years.
- The architectural complex was put into use again in June 2003, and has operated for over 1 year so far.
- The restoration of the Hall was completed with the associated efforts of the non-governmental organizations and the governmental sectors.
- Through the preservation and restoration of the architectural complex, the typical example of the local ancestral hall architecture in Foshan is hence preserved, so is the spirit of the place reflected by the Hall as the critical social space.
- The architectural complex, which is well incorporated with the open space, creates a key node at the traditional spatial development axis of the city.
- The restoration program adopted substantially the traditional local craftsmanship, including the rubbed brick with tight joints, brick carving, wood carving, stone carving and lime-straw sculptures on walls.
- The restoration involves reasonable uses of the traditional techniques, such as the traditional dampness prevention treatment, wood works and stone works craftsmanship, meanwhile engages traditional craftsmen to teach the young craftsmen who are determined to pursue the careers of historical architecture protection.
- Proper selection of building material, and the overall structure and local pattern represent the typical local and historic characteristics. The wood material are elaborately used (as reflected by the different tropical wood Borneo ironwood, rose wood, missanda, pineapple wood) at different parts of the architecture.
- The Hall, with its openness and commonality, ensures its positive contribution to the surrounding environment and local communities, hence realizes the favorable social benefits; and, as contrast to the surrounding traditional architectures which have been massively demolished to spare space for the new buildings, the Hall sets up a landmark of the city reflecting the rich historic connotation and unique traditional features.

Chicks to soften sunshine

Burnished bricks on the wall

The ash models before and after conservation

□ Appropriate use/adaptation

- **Changes made to building (repairs and additions), and relationship of new to old**

Repairs

The Hall was completely repaired, including structure and ornament. The whole work could be divided into woodwork, tilework, earthwork , brickwork, stonework, etc.

Additions

Some missed members and indispensable parts for structure or function were rebuilt, while the ornaments impossible to retrieve the original appearances meanwhile doing no harm to the structural safety were not recreated. Some interior furnishings were also added to cater for the needs of the exhibitions in the Museum.

Relationship of new to old

The newly added furnishings are all movable, and their styles are different from yet harmonious with the original architecture. The added parts always comply with the principle of reversibility.

As Stated in ICOMOS *Charter on the Built Vernacular Heritage,* alterations which legitimately respond to the demands of contemporary use should be effected by the introduction of materials which maintain a consistency of expression, appearance, texture and form throughout the structure and a consistency of building materials.

- **If restoration involved changing original use, rationale for changing use**

The architectural complex used to be an ancestral hall, while now is used as the Cantonese Opera Museum after the restoration.

Rationale for changing use

Since the posterities of the Huang Family all migrated and settled down in Hong Kong, Macao and overseas, the Hall had been left at an abandoned and unused status, and could hardly serve the originally assumed functions. Also given the preconditions of not changing the building typology and appearance, ensuring the structural safety and not adding any fixed facilities onto the architecture, the changed public uses of the Hall as a pubic building to display Foshan's brilliant local cultural traditions without bring about any physical or biological destruction to the architectural complex.

Folk opera show

See through the chicks

Opera stage

South lane before and after conservation

□ Interpretation of the significance of the building(s) in the conservation work

- **Restoration methodology**

1. The paramount principle is to keep the original and genuine feature of the cultural relics, i.e not increase the member typology or fixed facilities; all technical and safety measures shall not change the original appearances of the relics.
2. It was determined through studies of due prudence to clear away and remove the unreasonable additions and alterations by the later generations to keep the original architectural style. In view of the fact that such additions and alterations had occurred in many places, it was through careful distinguishing and studying process and after finding out the specific changed parts according to the material, craftsmanship, furnishing style and the slightest traces remained, the redundant unreasonable additions and alterations were properly removed.
3. Principles of_materials: the part of the architecture that can be easily researched about and restored, or the parts that are influential to the normal uses or to the significance of the architecture would be repaired and restored in a way to keep the original members and materials as far as possible and to restore the old architecture as it had been. For original articles that must be replaced, the same kind of material must be applied; for original members that could not be reused, the new members would be supplemented, repaired and restored, meanwhile differentiated from the original material.
4. It was resolved not to restore the missing members that are impossible to be researched about and not affecting the normal uses of the architecture. For example, the completely missing ornaments on main ridges are not restored, while their traces remained and the members are carefully preserved.
5. Eliminating potential hazards and prolonging the service years of the architecture: it was resolved to eliminate or minimize the various potential hazards, and conduct all-around examination and improvement over the technical systems of the Hall, including the structure, drainage, fire prevention and lightening prevention etc, hence avoid the occurrence of any disaster, or maintain certain capability and resistance to deal with the emergency in case of any disaster.
6. It was resolved to clearly and accurately number all the rooms, courtyards, structural members, windows, doors and other freestanding members, and set up detailed technical archives.

Brick-carving on the wall

Woodcarving

□ Understanding of technical issues of conservation in the interpretation

- **Issues of conservation/restoration considered and justifications for solutions**

Problems encountered during the Restoration and Rationality of countermeasures

As stated in *Nara Document on Authenticity*, "Conservation of cultural heritage in all its forms and historical periods is rooted in the values attributed to the heritage. Our ability to understand these values depends, in part, on the degree to which information sources about these values may be understood as credible or truthful."

1. Overall architectural layout

The internal architectural layout was restored, with the unreasonable additions and alterations removed and the necessary access roads and courtyard restored. The stuff used to close up or block up the doors toward the outside were removed, and replaced with door-leaf for easy management.

2. Needs for Structural Safety and Normal Uses

- *Roof Woodwork and Tilework*

Owning to manmade destructions and lacks of repairs and maintenance for long time, the roofs were observed as dilapidated, rotten, decayed with rain leakage. Judged form the existing conditions, there were several types of roof damages.

- *Wall Surface Brickwork and Fenestration*

The damaged walls at the Main Gate were all restored in the original methodology, with the exterior wall surfaces all using the rubbed bricks with tight joints. The exterior wall surface used the rubbed bricks with tight joints; all blue bricks used here were the original materials, the portion of bricks short of supply were collected from the general public. The newly rubbed blue bricks were in tunes distinguishable from the original ones.

With the wall surfaces of the main architecture in the all courtyards being all rubbed blue bricks with tight joints, and the less important architectural part being plain brick walls, the architectures were provided favorable appearance effects and physical performance, and any additional plastering would adversely damage the original style of the architecture. So the existing plastering was carefully removed and the wall surface cleaned and washed to restore the original appearances.

The various holes were filled and sealed, and restored in the original methodology.

- *Earthwork and Pavement*

Totally there were 3 types of interior floor pavements : gray quarry tiles (360mm*360mm) pavement in an inclined 45° at various attics; the 20mm*20mm encaustic tiles pavement at Altar Pavilion; the red tile (470mm*470mm) pavement of offset joints at F1 and F2 of the back courtyard. The pavement of the halls would be renewed in accordance with the original methodology, i.e. rebuild the bed course, replenish the fluvial sand and repave the quarry tiles. The Altar Pavilion was paved with original encaustic tiles; the middle part of the original pavement, which was substantially sunken and broken at several places, was repaved with the original material; the bed course was also replaced to facilitate the drainage.

• *Wood Attics*

There were 7 wood attics in the Hall respectively at A1, A3, C1, C3, G1,G2 and G3, of which G1 and G3 were added after 1949 featuring obvious traces such as inferior materials, non-uniform specifications and simple techniques. The original structures of the attics were regular, even and finely fabricated rectangular thick ribs, with bamboo strips to connect the wooden planks and veneering of quarry tiles of diagonal edge joint.

• *Column*

The structural columns include wood columns and stone columns, totaling 22 wood ones and 14 stone ones.

The wood columns were in satisfactory conditions, however, the column bases were extensively cracked; moreover, many were eaten by termites to various extents including two most seriously damaged. The one most seriously damaged (MZ14) was reinforced with hoop iron at the base, and all cracks were treated and filled up with pouring of epoxide resin.

The stone columns, except for the one broken at the Main Gate, were kept basically intact, so a new column was chiseled in accordance with the original pattern still using the original column base.

• *Beam Frame*

The beam frames were all made of belian ulin, and were kept basically intact except for the one seriously damaged at the Main Gate. Some parts were attached by termite and some parts were altered. One added bracing was noticed for the beam frame of the southern subcentral bay of G2.East part of the Main Gate was completely rebuilt with the original materials and members to ensure the safe construction and structure. With regard to the part with added purlins, the redundant members were removed, meanwhile replaced with original members for consistent material and color effects.The main beam and some purlins of the Back Mansion seriously damaged by termite attacks were replaced respectively with camphor wood main beam and cedar wood purlins.

3. Material Selection and Sourcing

The removed parts of the architecture were rebuilt with the original members, and numbered and configured by referring to the symmetrical side of the beam frame, all using connection of tenons and mortises. When the original members could not meet the demands, the same would be sourced from the general public, or the same material would be used to remake the members in the original patterns.

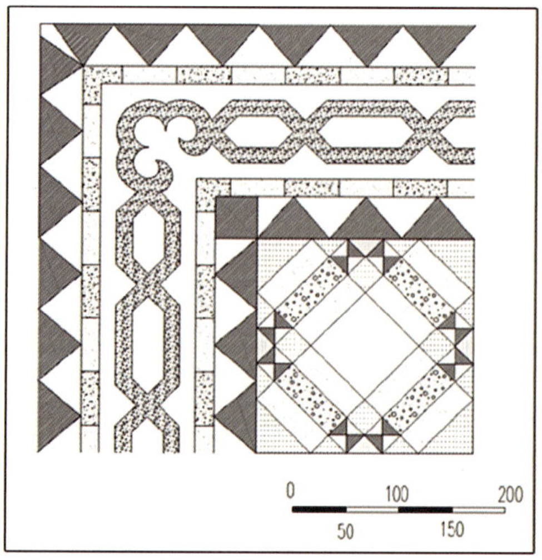

Detail of flooring pavement

Brick-carving kept the previous condition

Corner of roof during restoration

4. **Pattern, Building and Ornamental Craftsmanship**

Both the building and the ornaments strictly follow the traditional techniques and methods.

a. Wood Member Ornaments

- The woodcarvings axed and chiseled during the Great Cultural Revolution would be kept to show damaged looks, hence accommodate the traces of various historical periods.
- Totally five wood que-ti were missing in the whole architectural complex. Considering the fact that que-ti served supportively in the structural system, and each pair of que-ti were completely symmetric, they were restored according to the appearances the counterparts on the opposite side.
- Eaves plates were basically preserved with ornaments destroyed to different degrees. The missing eaves plates were reinstalled, while the damaged ornaments on the plates were no longer restored owning to the complexity of the craftsmanship.
- The door-god on the rose wood main door was traceable hence was repainted with colors.

b. Stone Member Ornaments

- Stone beams, wood bearing blocks, handrails and balustrade posts were basically kept intact. The several places with missing articles were repaired with members of granitic plastering. Some had been damaged, displaced, or buried by brick walls. So the redundant brick walls were removed and the stone members were sorted. The missing parts were no longer resorted.

c. Other Ornaments

- The main ridge was no longer restored, while the ridge end ornaments were restored by referring to the one kept intact at ridge end. The grass dragon design had been seriously damaged to an extent to affect the overall appearances, hence was also resorted properly. The small red gable were also restored to original colors, while the damaged lime-straw sculptures and gable springer were no longer restored. The interior pothooks were kept to hang the couplet on the columns of the hall and the horizontal tablet with inscriptions. Since the door and window eaves ornaments, door heads and window heads could be ascertained through researches and the craftsmanship was well known, the original door heads and window heads were restored to the original patterns.

5. Technical Installations

The original drainage system had included iron roof gutter, iron down spout, aboveground open drain and sub-soil drain of stormwater gathering opening (made of stone or iron), which had been comprehensively designed yet without effective protection. Since the drainage system was still anticipated to serve its due functions, the restoration managed to restore all drainage members, improve the drainage gradient and dredge up the drainpipes and ditches.

6. Disaster Prevention and Control

a. Fire Control

Considering that the Hall had been primarily single-storied and only about 70m in length with open surroundings, it was resolved to adopt a fire prevention system comprising vehicle fire hydrant, hose fire hydrant and fire extinguisher. Meanwhile, each courtyard was installed with two foam fire extinguishers, which were placed on the wooden stand at the corner instead of wall hanging so as to avoid any damage to the walls.

b. Lightning Protection

With a view to avoiding any damages over the original images of the architectural complex, iron sheets were paved on the main ridges of the main architecture and plastered with cement upside, then lay the reinforcing steel bars along the inside of the side ridges and connect with the ground along the exterior walls in the west side.

c. Termite Prevention

After the treatment by the Termite Prevention Office of Kaiping City, the once rampant termite in the Hall was under effective control, and no new traced were observed so far. Since then the Hall has been examined regularly (once a month) about the status of termite.

d. Other Disaster Prevention and Control

Since the Hall is sheltered from many tall buildings in the surroundings, and there is few protruding element on the roof, it is unnecessary to take special windproof measures. The architectural complex primarily in single-story brick-and-timber construction, after its structural improvements, does not need any special earthquake prevention measures.

☐ Use of appropriate building, artisan, and conservation techniques

Building techniques: brick rubbing, fenestration, lightning proofing; wall masonry include rowlock cavity wall, all heading bond, or one stretching bond every *N* heading bonds.

Technical means: manual painting; stone carving; woodcarving; mixture ratio.

Repair techniques: hoist with hinge and pulley; white ant prevention; deviation rectification; antiseptic treatment.

The conservation project engaged the local carving craftsmen with profound knowledge about the traditional craftsmanship from Gaoming of this area, as well as the skilled technicians from Wuhua County specializing in the ancient architecture conservation.

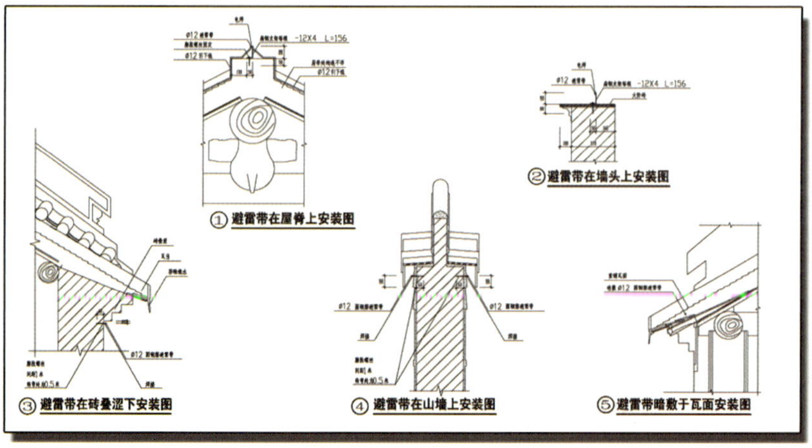

□Use of appropriate materials

Restoration materials used include:

(1) Blue brick, used for wall masonry and furnishing, specifications:
a.253x110x57, b. 253x110x53, c.254x110x57, d.257x110x61, e. 262x110x62, f.280x110x62.

(2) Granite, for exterior pavement; bluestones were also used for stone carvings.

(3) Larger tile, including two specifications: 470mmX470mmX10mm, red, for interior pavement of the first floor; 360mmX360mmX6mm, gray, for floor pavement in diagonal pattern.

(4) Encaustic tile, for pavement of the Altar Pavilion.

(5) Roofing tile: local flat tile.

(6) Wood: The belians were used for wood columns, arches and purlins, the pineapple wood for que-ti column brackets, the teak wood for hanging fascia and woodcarvings, the rose wood for the main gates. The other added member materials were sourced from the second-hand market.

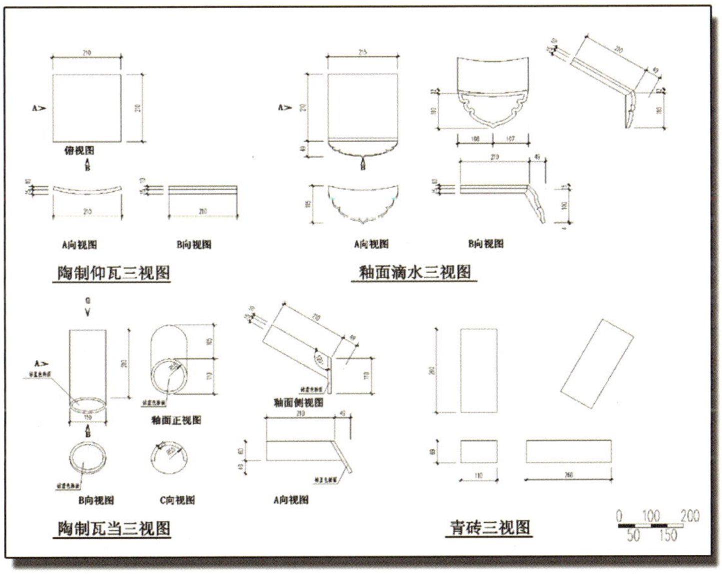

Details of Bricks and Terra-cotta tiles

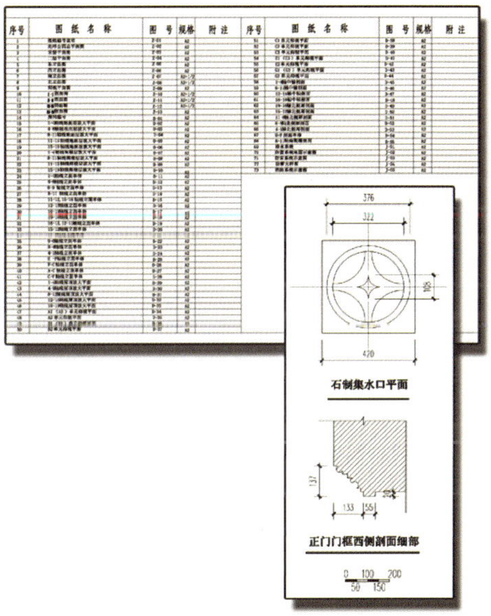

Detail of Stone members

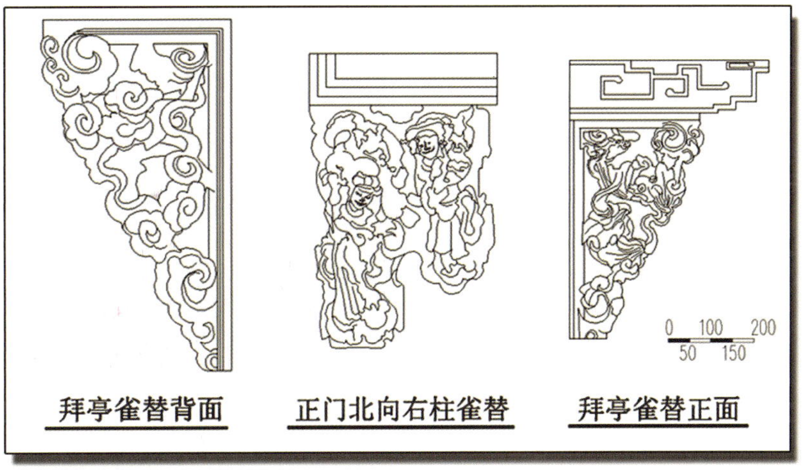

Details of Woodcarvings

□How well new elements and creative technical solutions respect building's character

The rationality of technical measures: The restoration efforts did not incorporate new variety of members. Furthermore, the various technical and safety measures would not bring about any changes to the original appearance of the cultural relics; the scaffold would not have direct contacts with the architecture; the fire control equipments were placed on separate stands instead of fixing onto the walls; the redundant members would be removed in a way not to enlarge the damaged area, and so on.

□Contribution to community's cultural continuum

- *Key stakeholders and involvement of local community in project*

1. Researchers and designers involved: Pro. WU Qingzhou, FENG Shu, FENG Jiang, OU Jie, Zhou Yigang and XIAO Min.
   Responsible institution: Academy of Oriental Archiculture, South China University of Technology was engaged to plotting, literature studies, textual research, craftsmanship study, technical development, formulating repair and maintenance principles, chairing restoration design, as well as studies over the traditional painting and carving of the architecture, and artistic support of restoration.
2. Lead craftsman: Mr. Xie Weiguang and Mr. Deng Jicai; Institution involved: The First Building Construction Company in Wuhua County, Guangdong Province.
3. Non-governmental organization: Cantonese Opera Museum
   As a museum raising fund and collecting exhibits via nongovernmental channel, it substantially funded the restoration program and would be the lessee and user after the completion of restoration.
4. Governmental sectors: Foshan Urban Construction Bureau; Foshan Culture Affairs Bureau
   The functional department concerned of the government, with the rights of entrustment, was responsible for inviting experts and professionals for surveying, plotting, investigation and restoration designs, proceeding with construction bidding, maintenance organization, management and evaluation in accordance with legal formalities, and partially sharing the expenses incurred. Meanwhile, Foshan municipal government was responsible for coordinating the museum study, uses and management, and educating the general public on how to protect and maintain the local cultural relics.

Daily show by fans of local opera

Pro. Wu Qingzhou and researchers

Decorated vehicles in the Opera Cultural festival

□ Impact of project on conservation practice and policy

The conservation project
- formed an integral part of the park;
- created a venue to showcase the renowned historical city;
- promoted the cooperation of the governmental bodies and non-governmental bodies on cultural heritage protection;
- won extensive applause, the Hall was listed as the Provincial Historic Reservation Unit in 2002.

□Ongoing socio-economic viability and relevance of the project

• Regulatory framework of the project

1. International Documentation on Historical Heritage Conservation such as *Nara Document on Authenticity* , *Venice Charter* and *Charter on the Built Vernacular Heritage*
2. *Law of the People's Republic of China on Protection of Cultural Relics*

• Mobilization of financial and in-kind support from public and private sectors.

Funding: Cantonese Opera Museum; Foshan City Government.

Non-cash support from public and private agencies: Donations of artwork and showpieces from oversea Cantonese Opera stars, fans and organizations.

Effect on Conservation Planning of National Historic City of Foshan and Conservation Planning of Traditional City Centre

Entrance torii of Cantonese Opera Museum

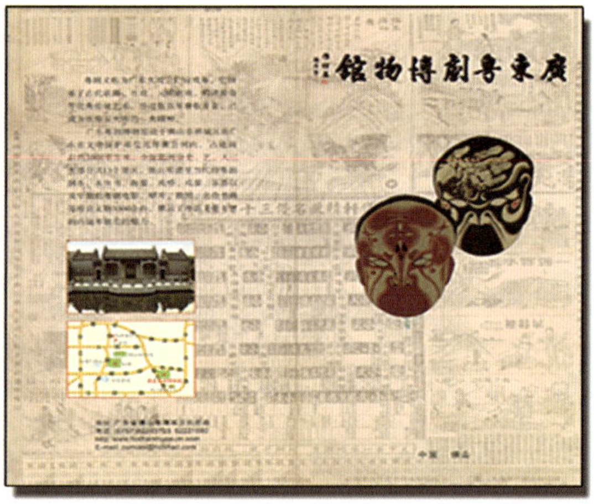

Brief introduction of Cantonese Opera Museum

- Contributions of each partner :
  Cantonese Opera Museum

Providing the majority of the funding amounting to USD459,770- in total, it is concurrently the lessee and user of the Hallafter the completion of restoration, and would provide exhibits and additional funding for routine management and maintenance.

Foshan Urban Construction Bureau, the coordinator of multilateral conference and provider of some restoration funding. It was engaged in coordination of multilateral conference meanwhile also provided some financial support (USD218,390).

Daily show of local opera

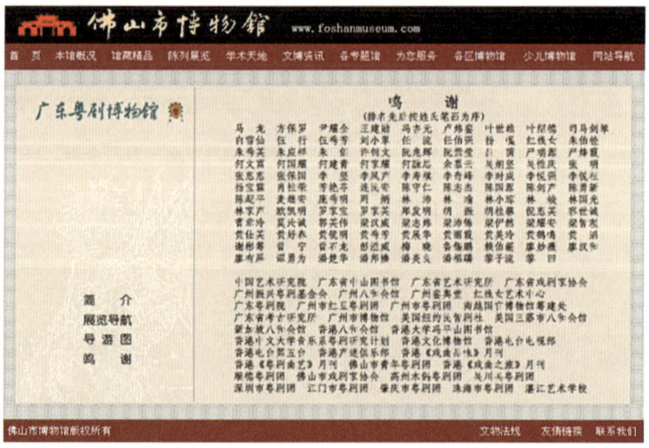

- *Provision for future use and maintenance*

Provision for future use: the Hall shall be used for common cultural purposes only such as visiting and exhibition, but not for commercial purposes.

Restriction to number of visitors each time: less than 10 persons for each attic. Restriction to annual visitor flow: no more than 200,000.

Provisions of maintenance: under the jurisdiction of the department in charge of cultural relics management; the daily maintenance frequency and working contents etc.

Regular examination of the Hall(once a month) for the termites status.

Publications and Concerned Medium Reports

□The complexity, sensitivity and technical consistency of the project methodology

With a view to catering for the space required by the modern community activities, the environmental design gave adequate consideration to the sensitivity of the historical heritage toward the measures of the project, specifically in the following aspects:

• *Historical Sense of Space and Scale*

The Hall was planned to use as a public museum, and the existing spatial layout then could realize this new use and function on the precondition of properly arranging the exhibits and visitor flows; hence the project showed adequate respects to the actuality of the architectural group.

• *Environmental Sensibility of Furnishings and Facilities*

As for the unavoidable newly added facilities of the modern uses and functions (such as exhibition, lighting and fire prevention), it was required to reduce their visual impacts over the historical environment meanwhile maintain the necessary recognizability. The project emphasized not to add any new furnishing onto the historical architectural group, and that the very few added furnishing worth displaying for the uses must be movable.

• *Technical Consistency*

As for measures of the conservation project to eradicate the potential safety hazards and restore the authenticity of the architectural group, the consistency of the technical means was accentuated, i.e. use the traditional materials and craftsmanship to complete all the works.

Construction Technologies

The construction sequence adopted by the project can be summarized as "handing over all technical information and details before the commencement of the construction, and referencing and learning from successful models and experiences".

The stone columns of 250×250×6000mm are successfully hoisted and installed in place once for all with combined efforts of the crosshatched wood supports and two hand-operated chain blocks. Workflow of the rubbed brick with tight joints includes: cutting into block, edge trimming, rough rubbing and fine rubbing. As the Hall uses bricks of 5 specifications, the finished products are put in good order as per the specifications, hence the bricks of appropriate specifications are easily available during the construction, and the wall surfaces thus built are of tight joints with the surface evenness and vertical degree both controlled within 3mm. In light of the maximum 7 layers of the roofs in the Hall, the inspection and maintenance on scaffold should properly check the way of their splicing, meanwhile make record in written form and with photography; the straightness of the row were controlled within 8mm and the tile surface evenness within 25mm. The three 6-meter-long girders in the hall of the last building row are replaced by using the wood support and the hand-operated chain blocks. Besides, new techniques were applied to an appropriate extent without damaging the interest, environment, layout and safety of the architectures.

• *Prevention of Atmospheric and Dust Pollution*

The wood machinery and brick cutting machines at site are often moisturized with watering to reduce the dust pollution.

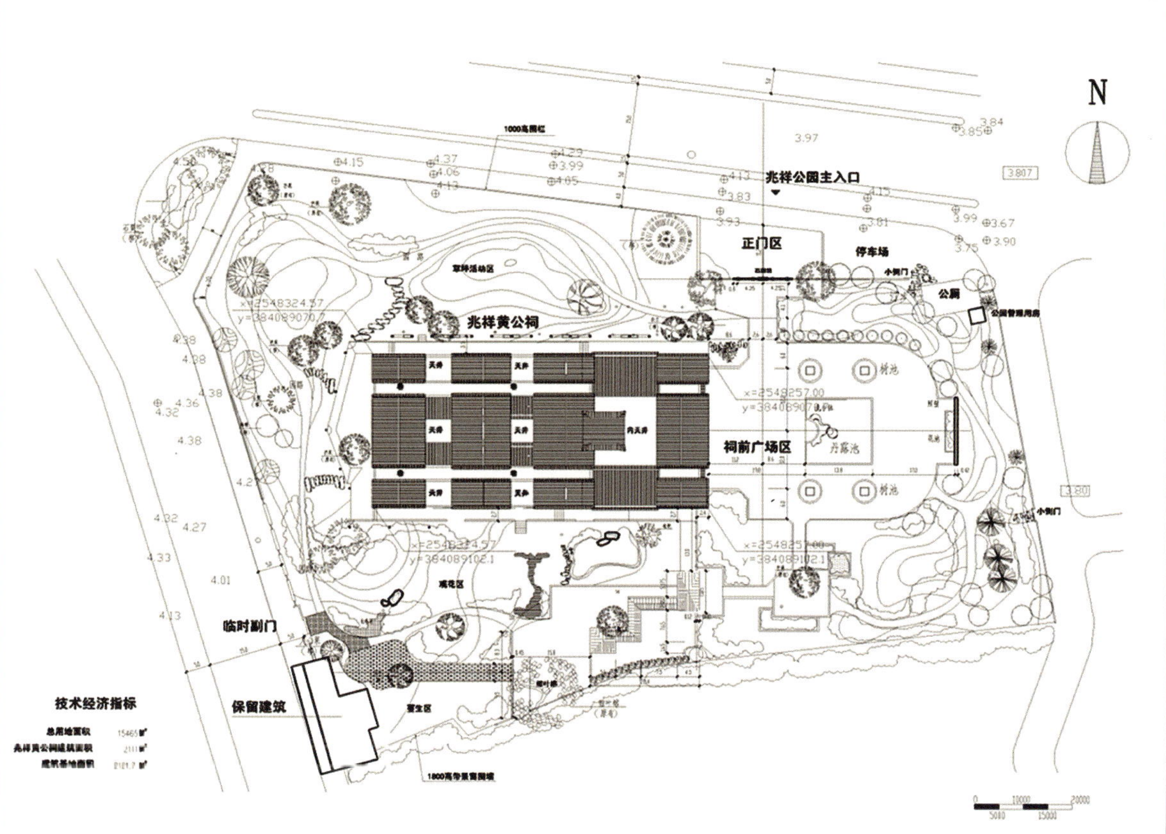

Site Plan of Zhaoxiang Park

The Hall now rises within the Zhaoxiang Park which is open to the general public for free. On its east lies a small plaza and a square pool, while the park's veranda and generous green space on the other sides. The Hall is currently used in good conditions.

It was resolved to clearly and accurately number all the rooms, courtyards, structural members, windows, doors and other freestanding members, and set up detailed technical archives.

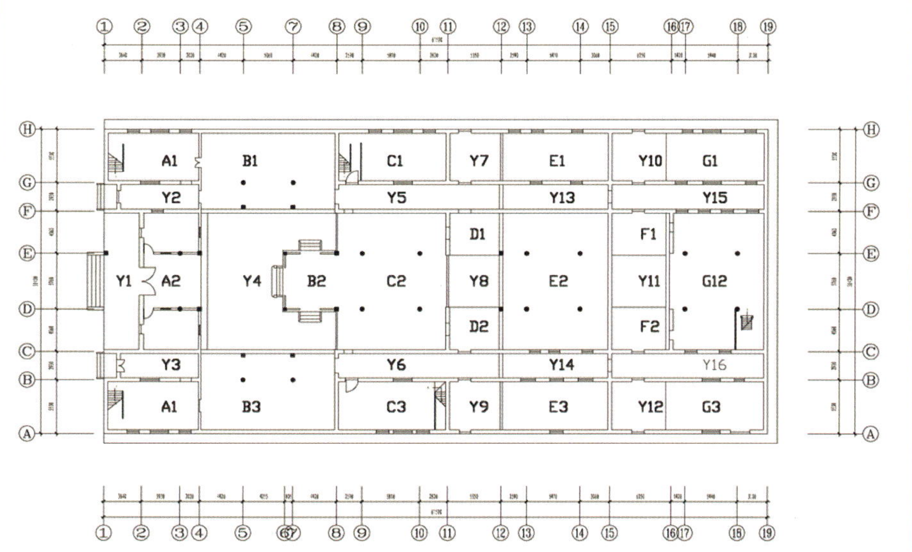

Number of Rooms and Courtyard

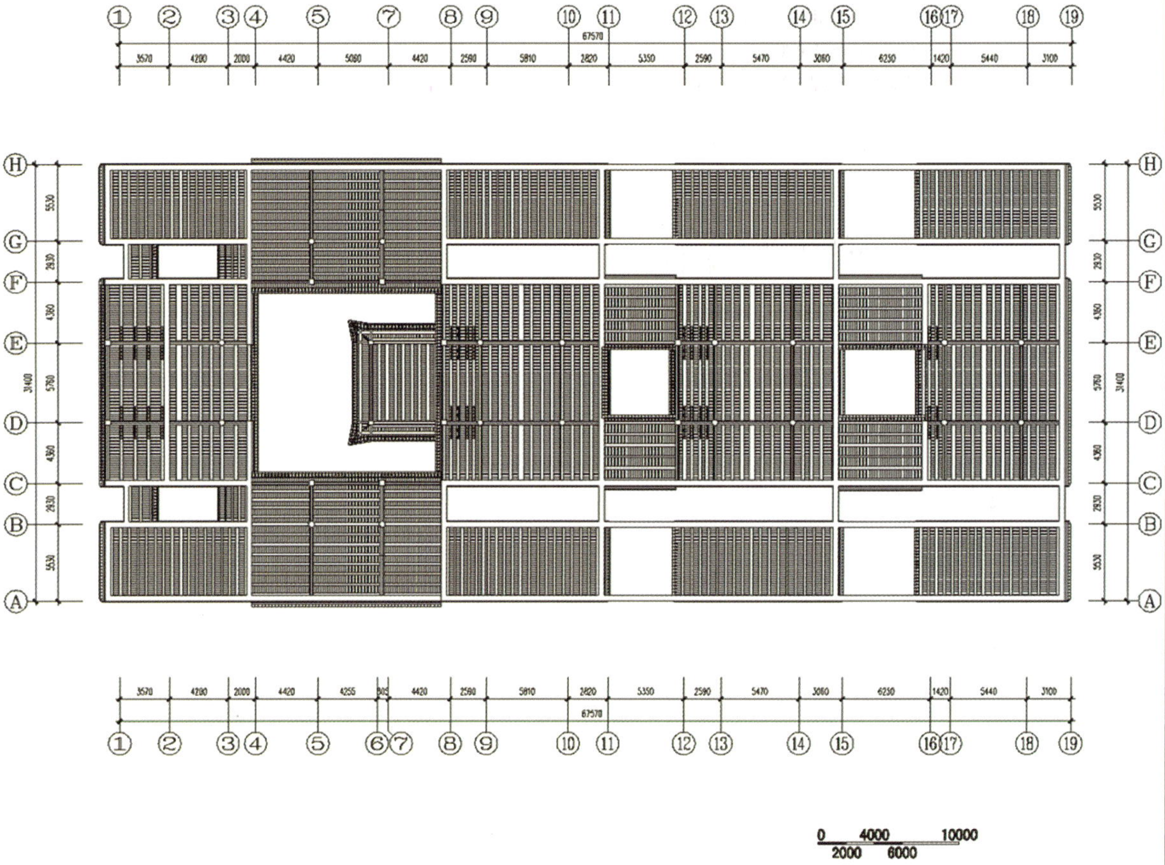

Roof Plan (Look up)

Positioned in the west and facing the east, the Temple covers a land area of about 3,060sqm, with the architectural group of 67.57meter in length and 31.40meter in width. The architecture program features courtyards of three sections, where the main architectures like the Main Gate and Alter Pavilion are all placed along the central axis, on both sides of which flank symmetrically the wing-rooms which are connected with the main architecture through laneways, showing the typical Ancestral Temple typology in the Pearl River Delta.

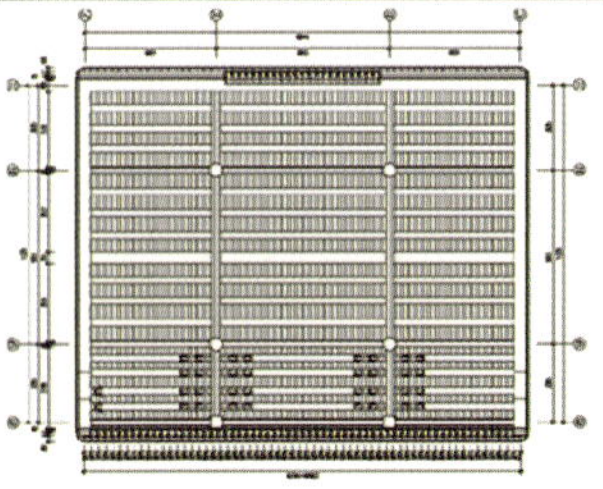

Roof Plan of C2

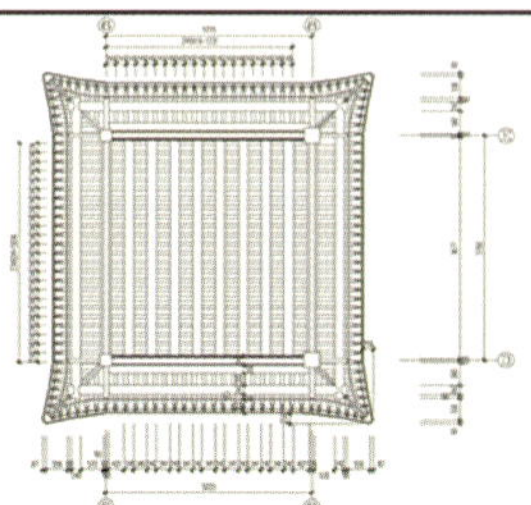

Roof Plan of B2

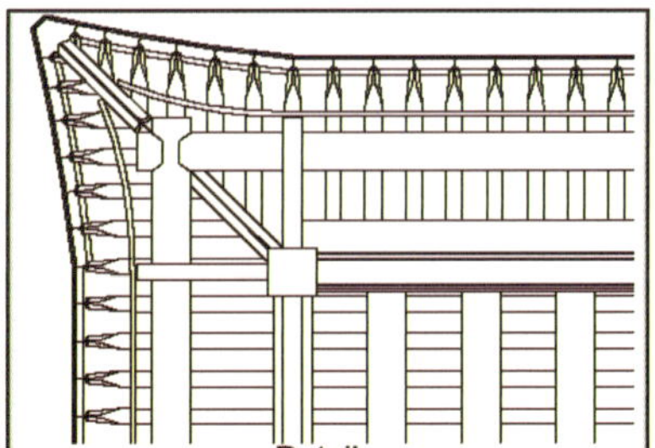

Details

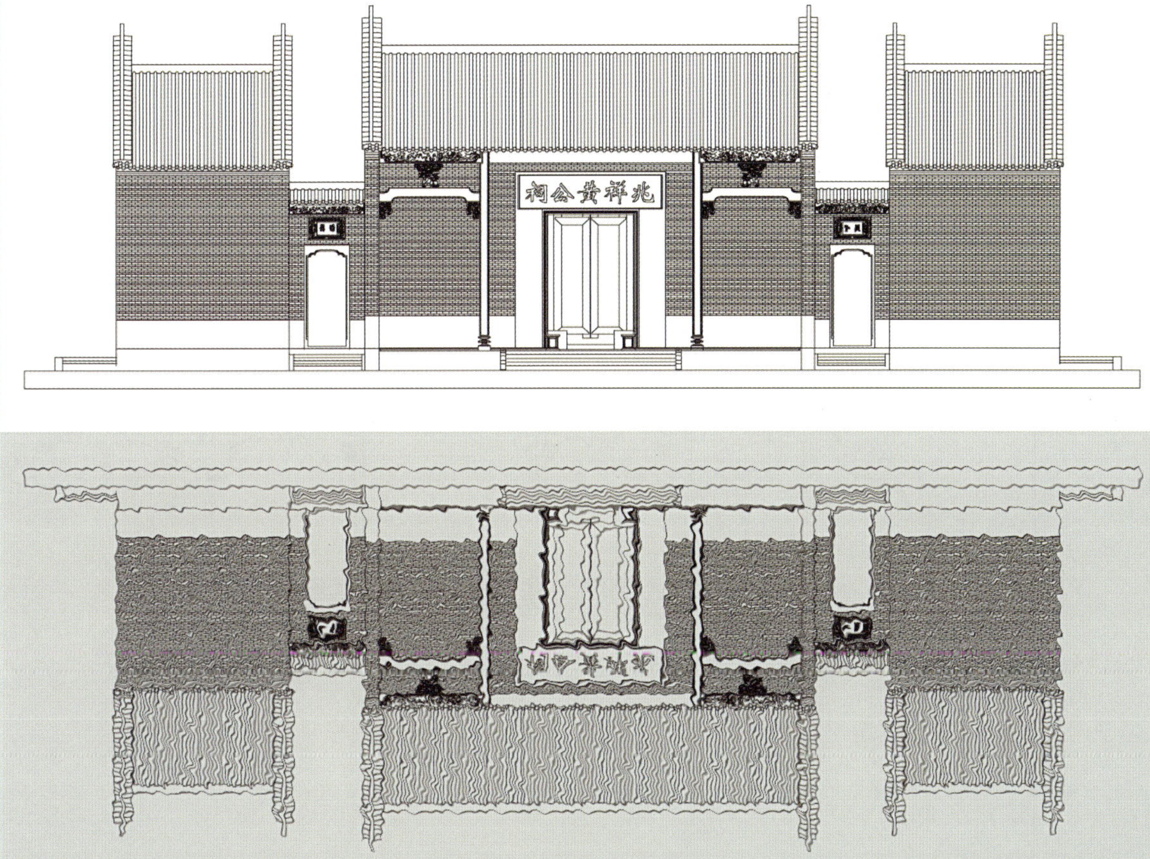

East Facade

The damaged walls at the Main Gate were all restored in the original methodology, with the exterior wall surfaces all using the rubbed bricks with tight joints. The exterior wall surface used the rubbed bricks with tight joints; all blue bricks used here were the original materials, the portion of bricks short of supply were collected from the general public. The newly rubbed blue bricks were in tunes distinguishable from the original ones.

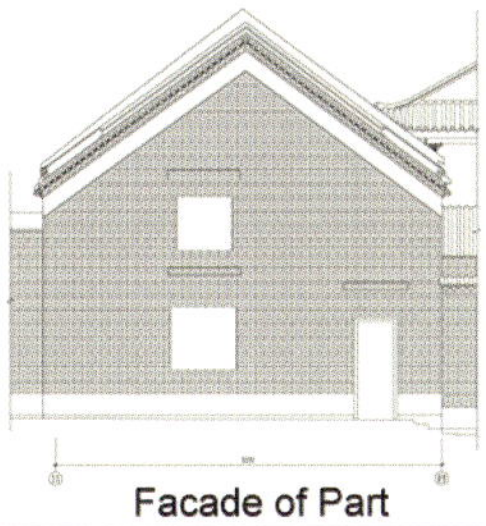

Facade of Part

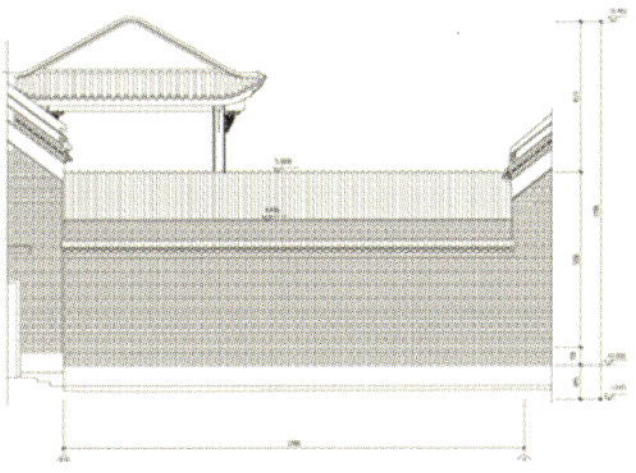

Facade of Part

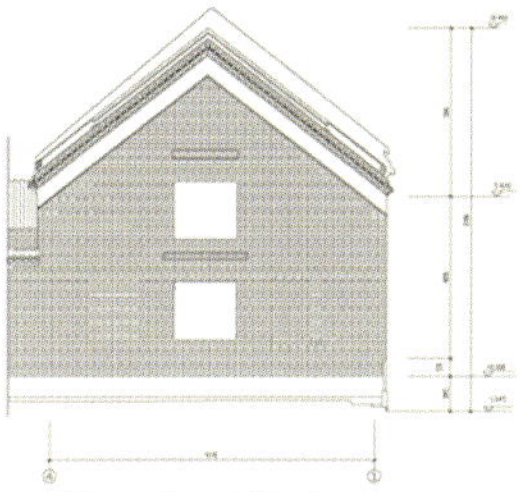

Facade of Part

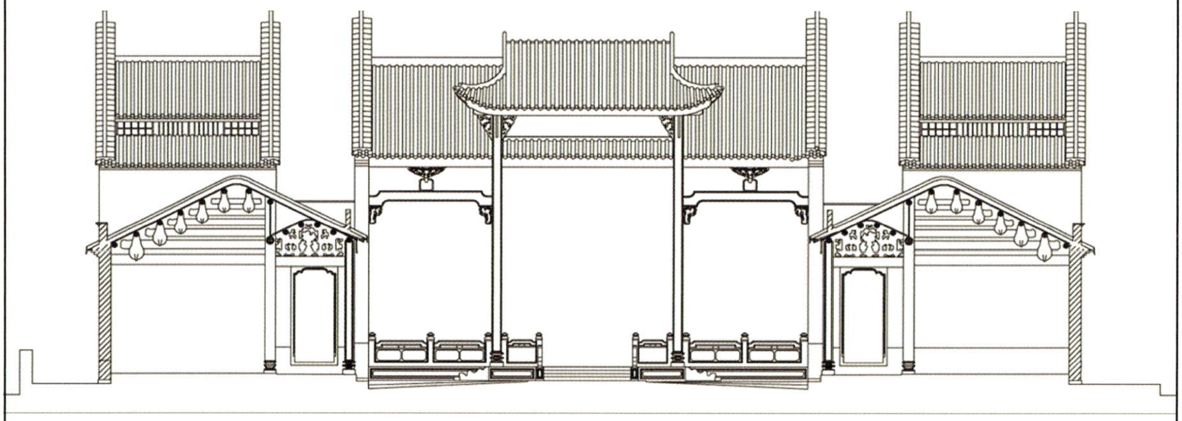

Section

The wood columns were in satisfactory conditions, however, the column bases were extensively cracked; moreover, many were eaten by termites to various extents including two most seriously damaged. The one most seriously damaged (MZ14) was reinforced with hoop iron at the base, and all cracks were treated and filled up with pouring of epoxide resin.

The stone columns, except for the one broken at the Main Gate, were kept basically intact, so a new column was chiseled in accordance with the original pattern still using the original column base.

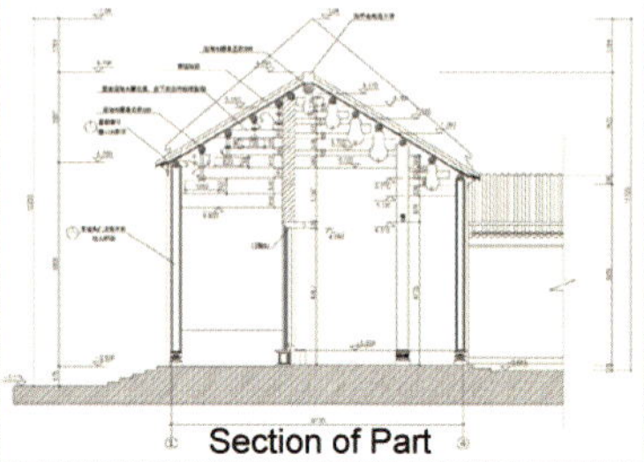

Section of Part

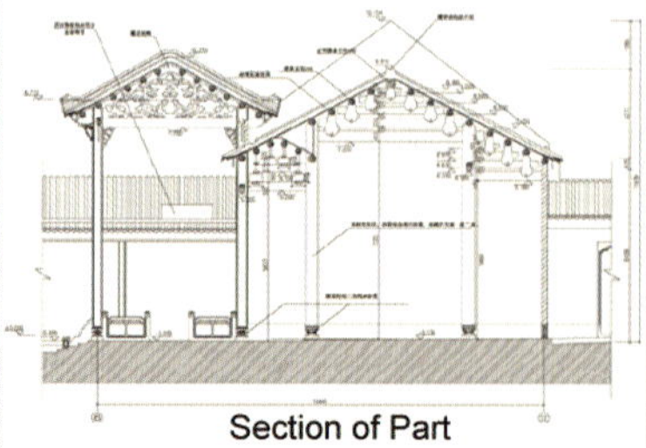

Section of Part

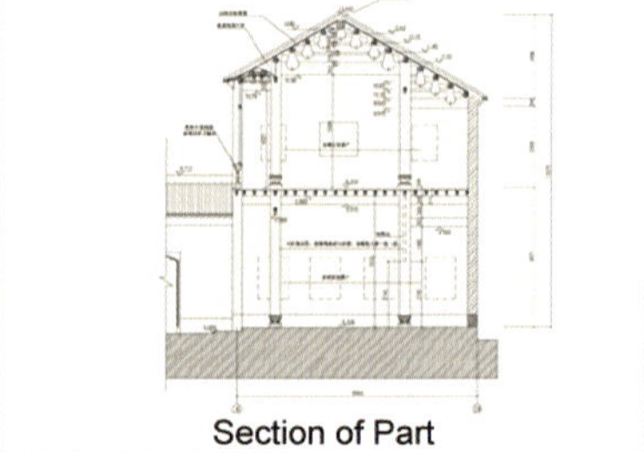

Section of Part

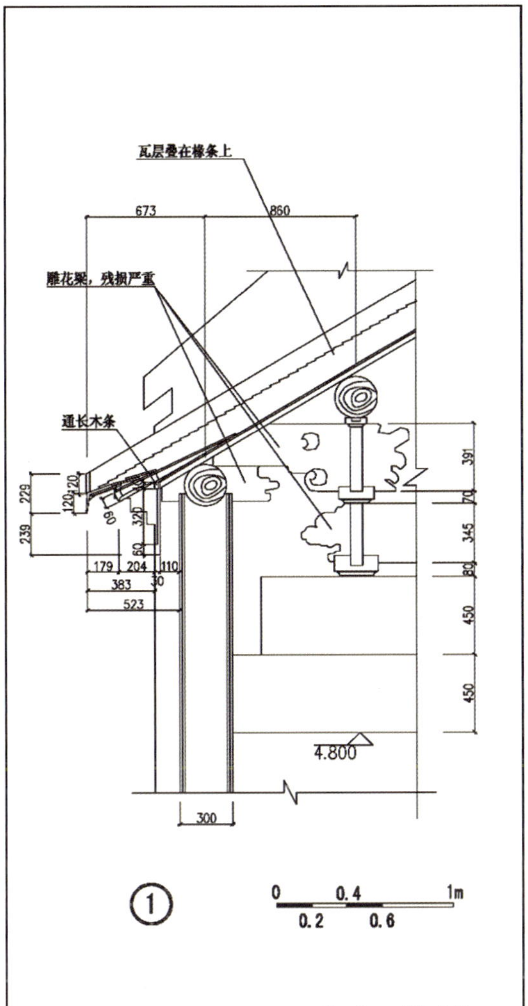

Section of Details

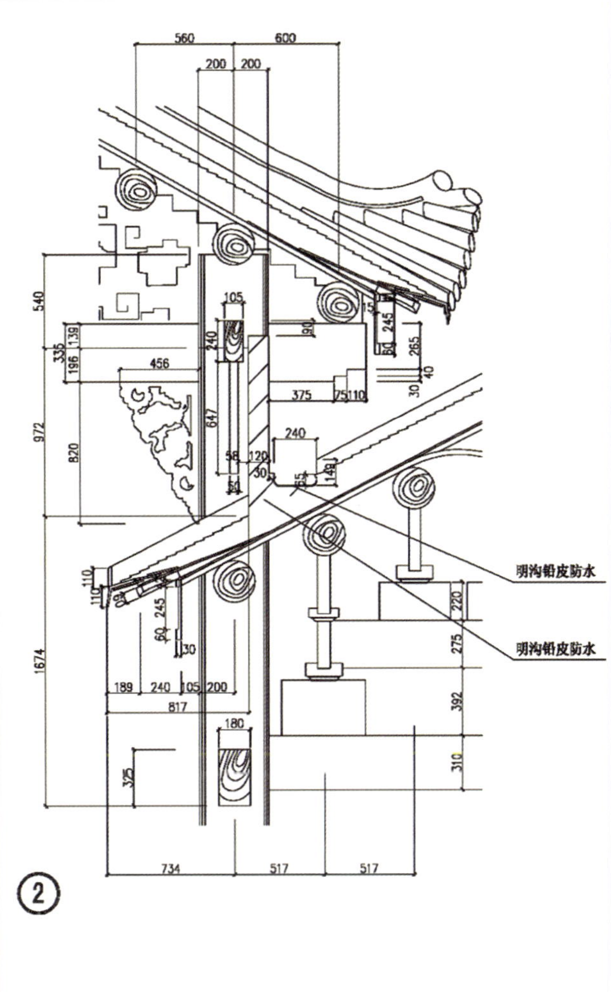

Section of Details

The beam frames were all made of belian ulin, and were kept basically intact except for the one seriously damaged at the Main Gate. Some parts were attached by termite and some parts were altered. One added bracing was noticed for the beam frame of the southern subcentral bay of G2.East part of the Main Gate was completely rebuilt with the original materials and members to ensure the safe construction and structure.With regard to the part with added purlins, the redundant members were removed, meanwhile replaced with original members for consistent material and color effects.The main beam and some purlins of the Back Mansion seriously damaged by termite attacks were replaced respectively with camphor wood main beam and cedar wood purlins.

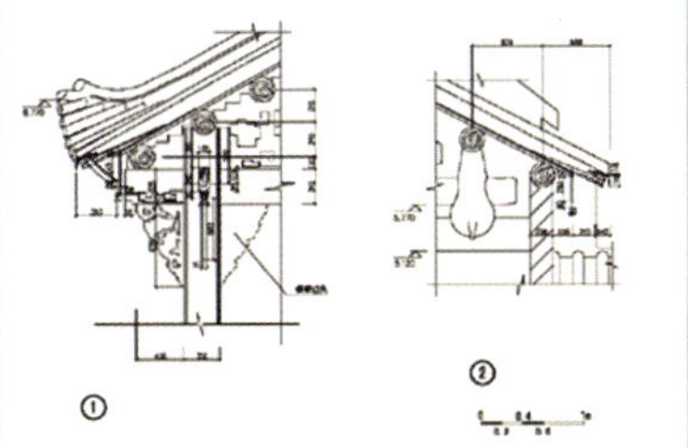

Section of Details

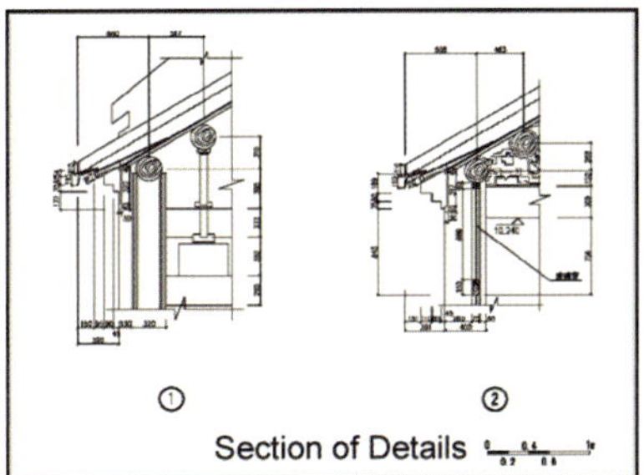

Section of Details

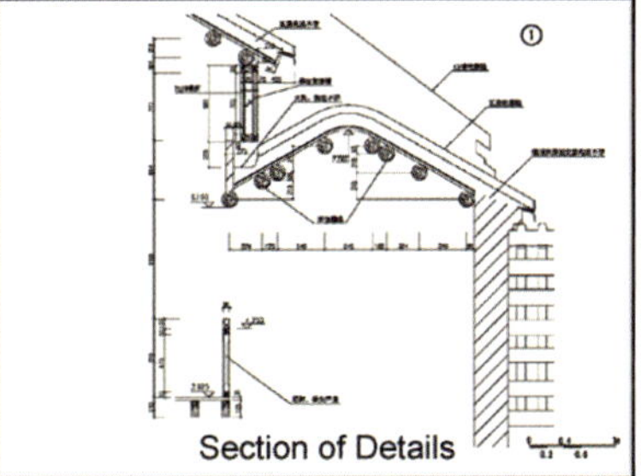

Section of Details

Details of Woodcarvings

The woodcarvings axed and chiseled during the Great Cultural Revolution would be kept to show damaged looks, hence accommodate the traces of various historical periods.

Totally five wood que-ti were missing in the whole architectural complex. Considering the fact that que-ti served supportively in the structural system, and each pair of que-ti were completely symmetric, they were restored according to the appearances the counterparts on the opposite side.

The door-god on the rose wood main door was traceable hence was repainted with colors.

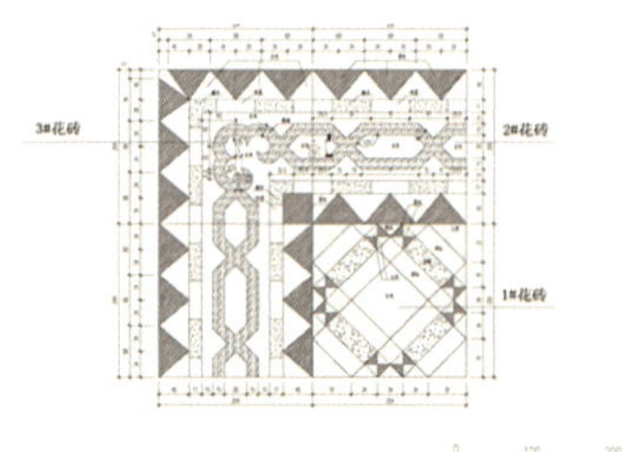

Details

Details

Details

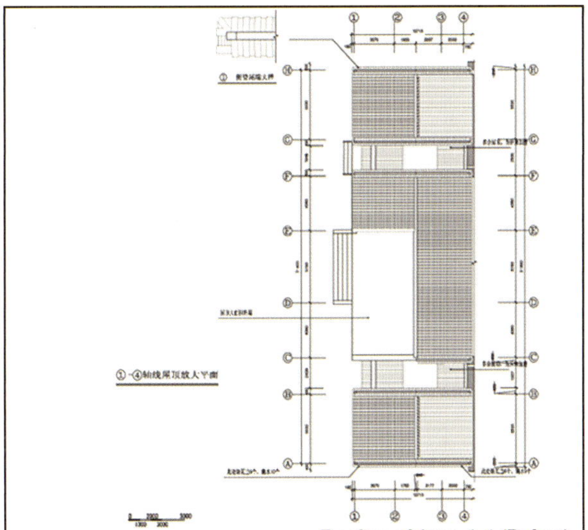
Rooftop of Axes 1-4 (Before)

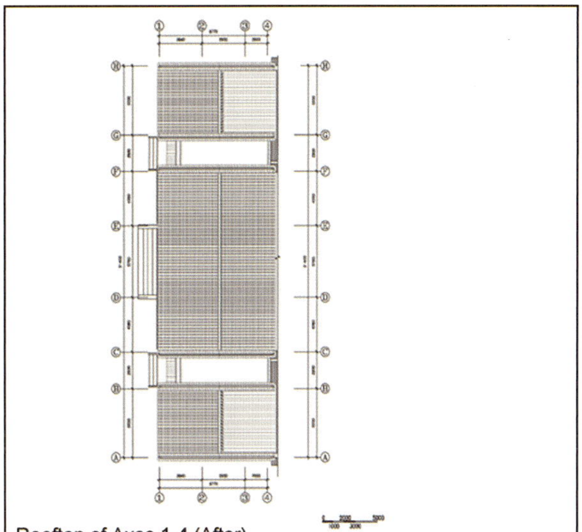
Rooftop of Axes 1-4 (After)

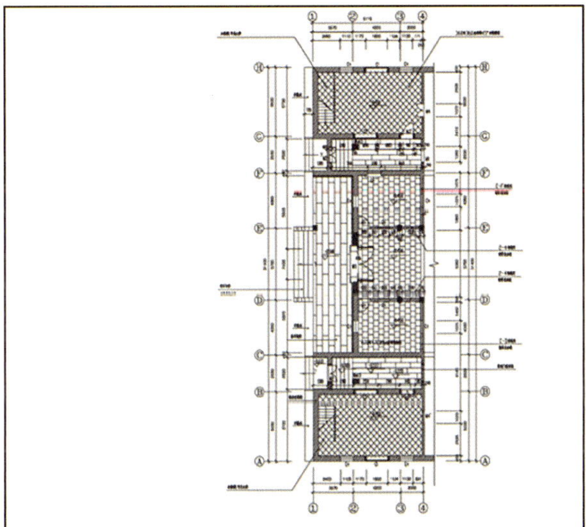
Site Plan of Axes 1-4 (Before)

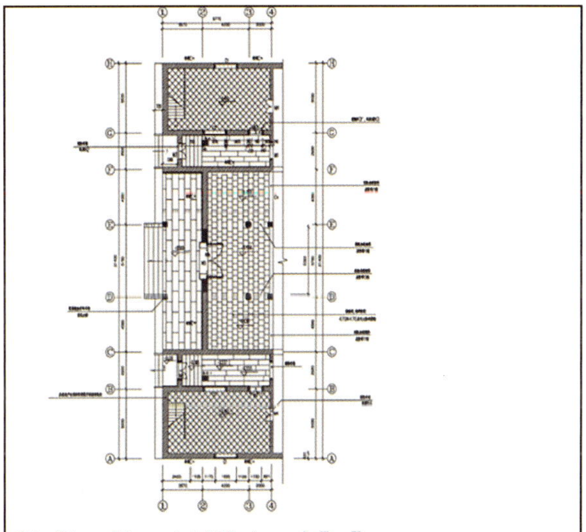
Site Plan of Axes 1-4 (After)

As per the common practice, the less important rooms with sloping roofs in north-south direction were be thickly paved with 36mm*36mm quarry tiles of offset joint, without pointed joint; whereas the main rooms with sloping roofs in east-west direction were thickly paved with 47mm*47mm quarry tiles of offset joints and edge joints. The Altar Pavilion was paved with original encaustic tiles; the middle part of the original pavement, which was substantially sunken and broken at several places, was repaved with the original material; the bed course was also replaced to facilitate the drainage.

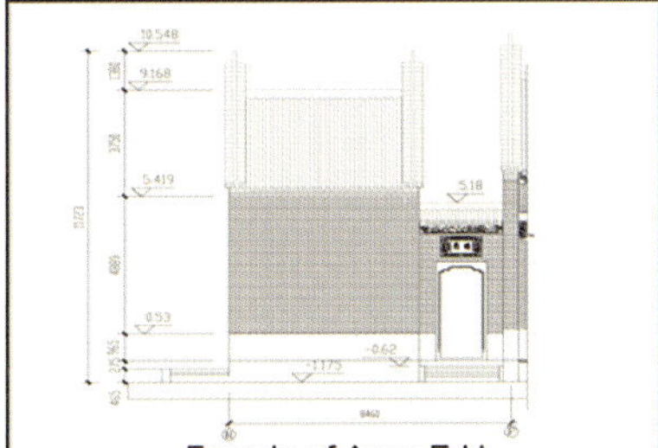
Facade of Axes F-H

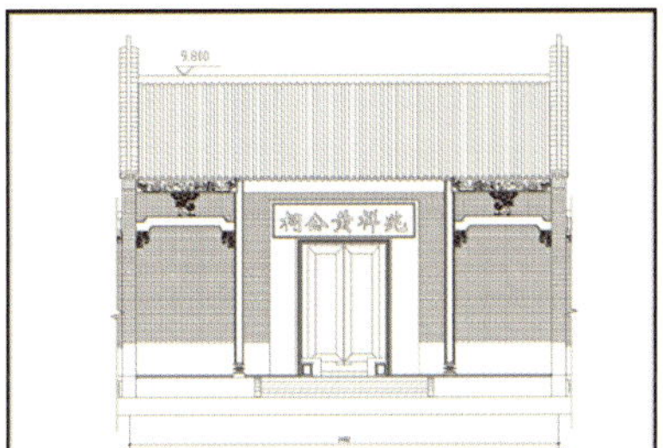
Facade of Axes C-F

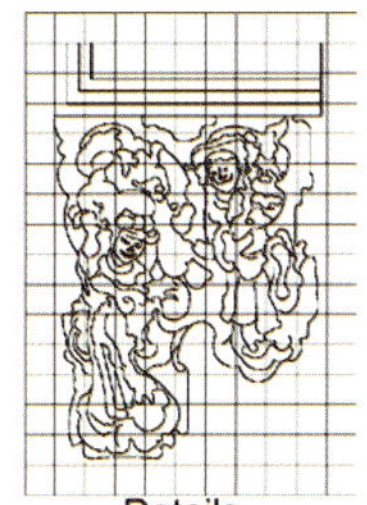
Details

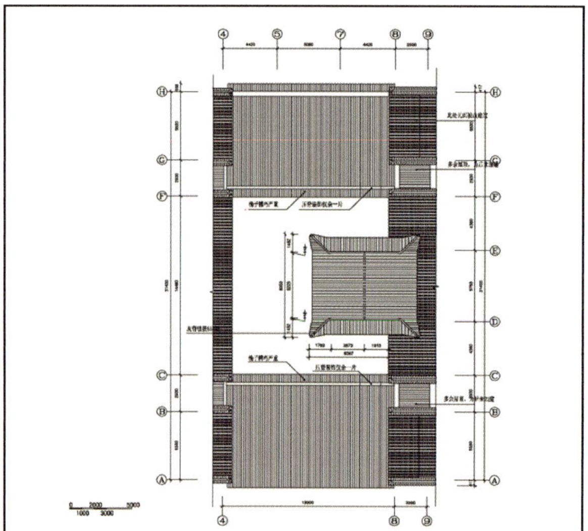

Rooftop of Axes 4-9 (Before)

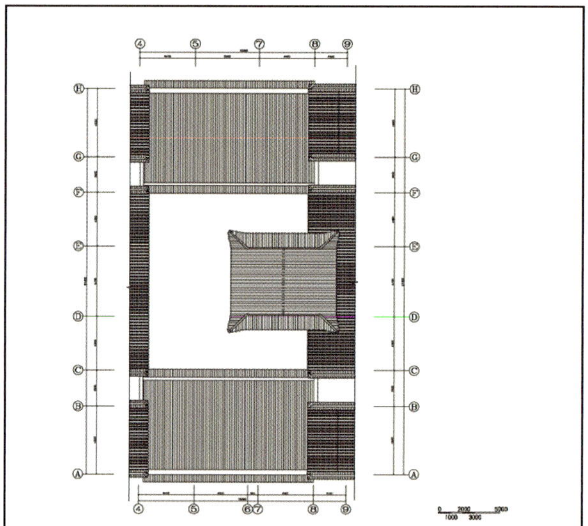

Rooftop of Axes 4-9 (After)

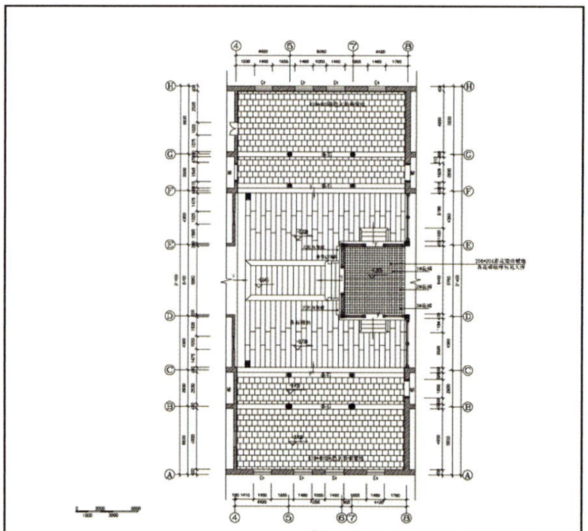

Site Plan of Axes 4-9 (Before)

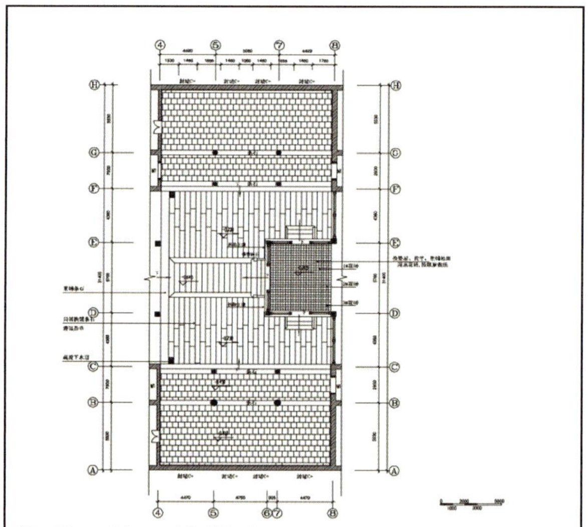

Site Plan of Axes 4-9 (After)

Various atria, courtyards, middle corridors and the platform outside the Main Gate were all paved with blue stone strips. Affected by constructions time and again, many places were sunken and cracked. These sunken places not only brought adversely influences to the general appearance and traffic, but also affect the water drainage and easily lead to waterlogging in the courtyards and alleyways. Hence all sunken and cracked areas were repaved, with the stone strips on alleyways paved much lower near the wall corners while higher in the middle part for natural ground drain purpose.

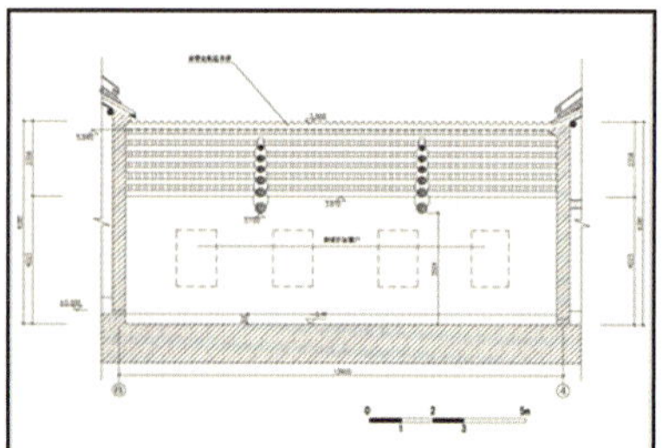

Section of Axes H-I

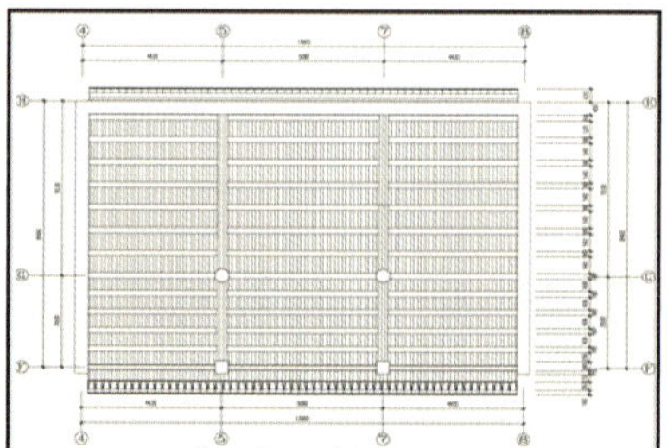

Section of Axes H-I

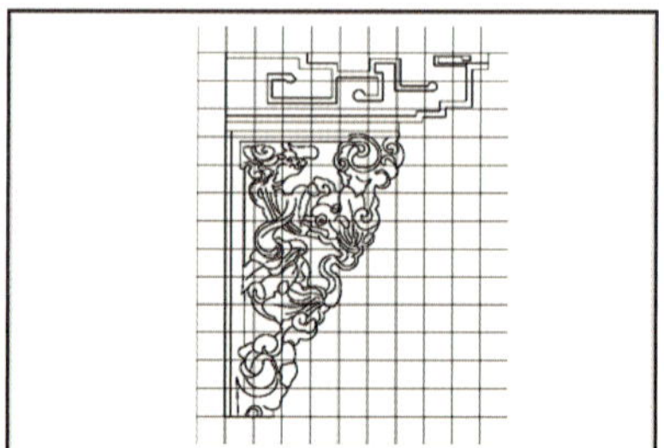

Details

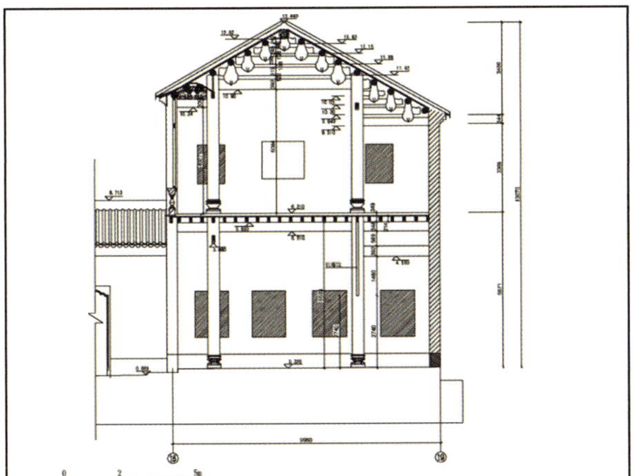

Section of Wood Attics (Before)

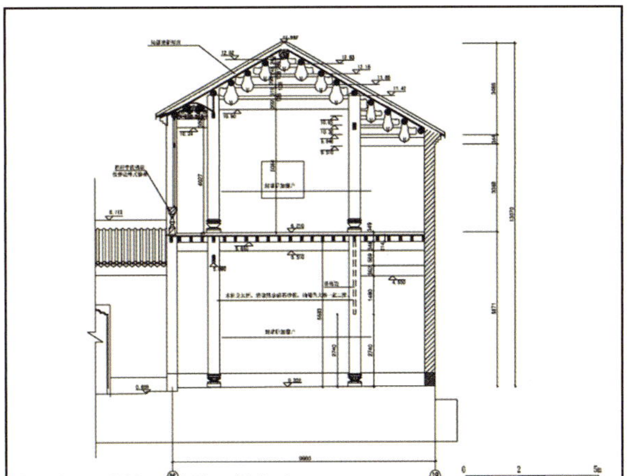

Section of Wood Attics (After)

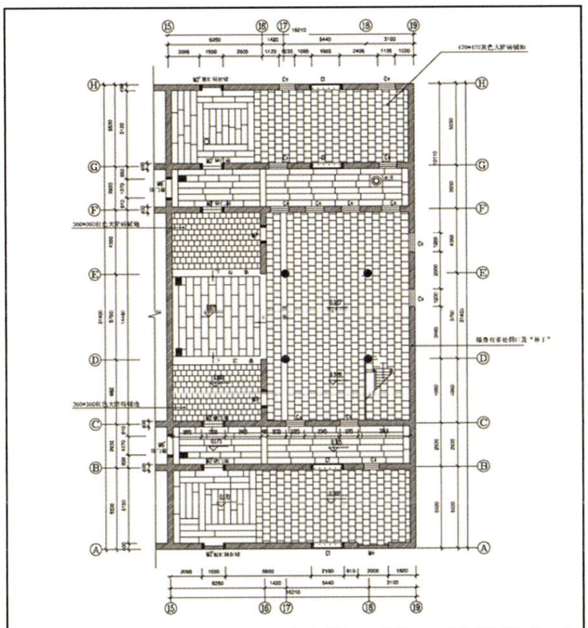

Site Plan of Axes 16-19 (Before)

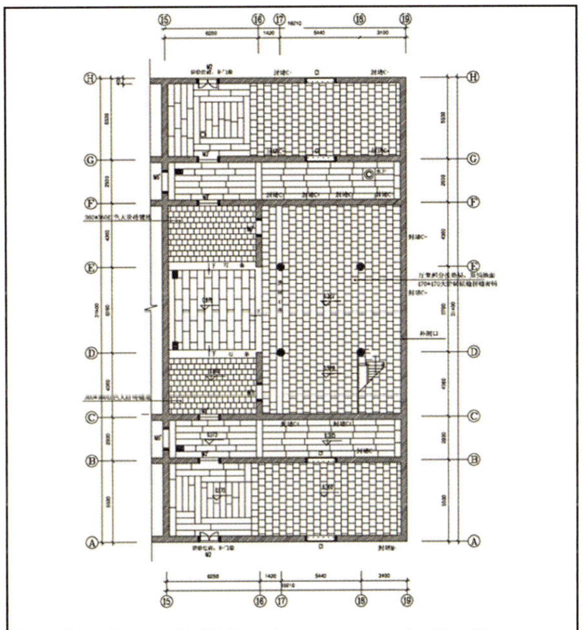

Site Plan of Axes 16-19 (After)

Attic at C1, which had been altered obviously with stairways placed at illogical position, was restored with reference to layout plan of C3. Moreover, G2 in the courtyard of the third section alongside the west wall had partially been hollowed out, and the added part thereto was distinguishable by either looking up at the ceiling from the ground floor or looking down at the ground from the attics. Such added parts were removed, and the hollowed floor was reinforced with handrails.

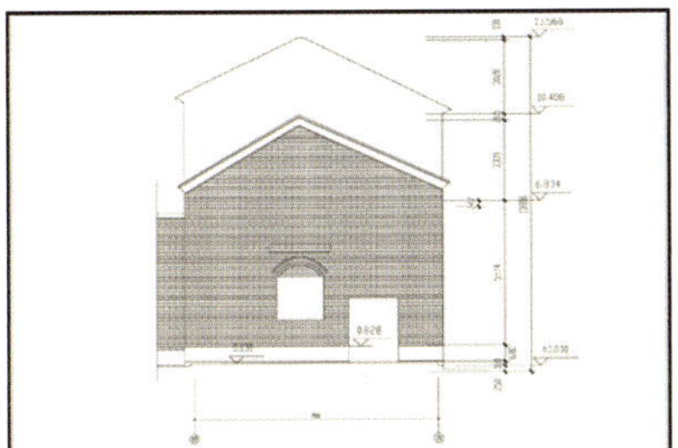

Section of Axes 16-19

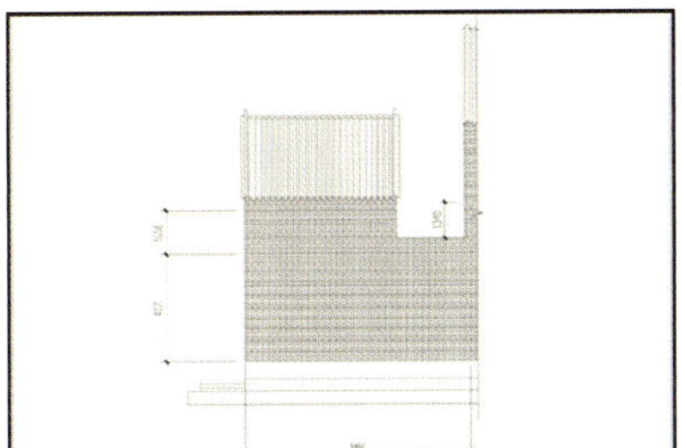

Section of Axes A-C

Details

CONSERVATION OF ZHAOXIANG HUANG ANCESTRAL HALL, Foshan, Guangdong, China

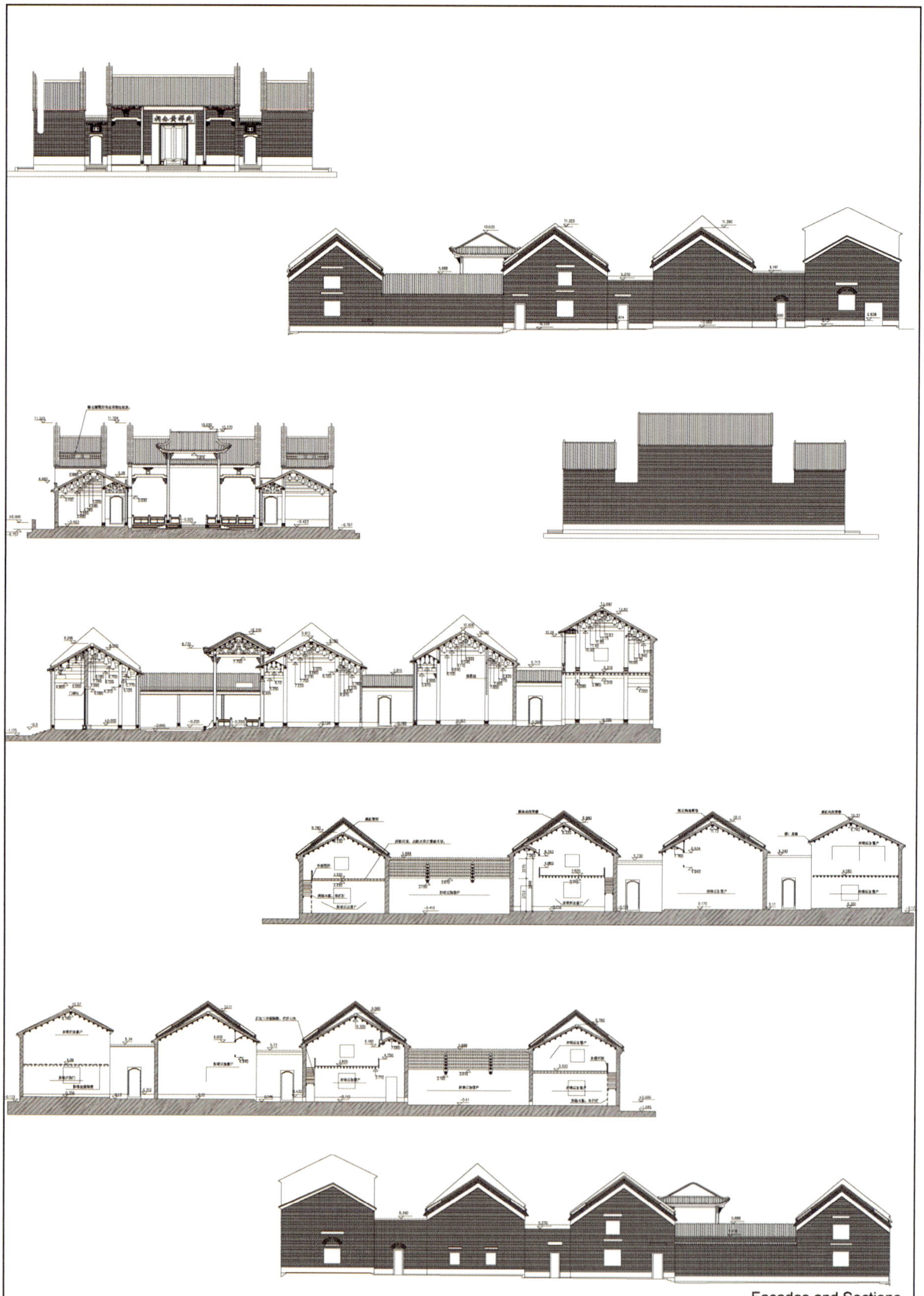

Facades and Sections

CONSERVATION OF ZHAOXIANG HUANG ANCESTRAL HALL, **Foshan, Guangdong, China**

UNESCO Asia-Pacific Heritage AWARDS for Culture Heritage Conservation

2005 UNESCO Heritage Awards for Culture Heritage Conservation

OCCUPANT COMMENT FORM

27 April, 2005

The **Conservation project of Zhaoxiang Huang Ancestral Hall** is being submitted for consideration for the 2005 UNESCO Asia-Pacific Heritage Awards. Please comment on your experience with this building, such as how well it works, how it relates to the community, etc.

Please return comments to **Pro. Wu Qingzhou** so it can be included in the entry materials.

Cantonese Opera is a kind of local opera with a long history and has the maximum fans all over the world. Zhaoxiang Huang Ancestral Hall offers a splendid exhibition space for Cantonese Opera.

While stepping into the courtyard through the main gate, you can feel as if there are actors vocalizing and acting. The lanes showing opera masks string the different exhibition rooms. Those naked blue bricks just make sense of the traditional culture, tell you the stories of past, and makes those superstars on view looked very kindness.

The visitors can be edified by the traditional architectural cultural when they appreciate the attraction of Cantonese Opera.

兆祥黄公祠古建筑为广东粤剧这一具有悠久历史的传统剧种提供了最好的展示空间。

一进入祠堂大门到中庭，仿佛进入一个古时的戏曲科班院落，能够听见角儿们在吊嗓；挂满脸谱的青云巷道纵深直入，串起各个展厅，墙面那些裸露的砖块，就像寻常人家的庭院。展厅展示的虽是那些人人仰望、风华绝代的红角儿，却使人感到他们很亲切寻常。

在参观我馆过程中不仅领略了粤剧文化的艺术魅力，同时也在受到了岭南传统建筑文化的熏陶。

Xiao Haiming

Cantonese Opera Museum

Zhaoxiang Park , Foshan , Guangdong , P.R.China

UNESCO Asia-Pacific Heritage AWARDS for Culture Heritage Conservation

2005 UNESCO Heritage Awards for Culture Heritage Conservation

OWNER CONSENT FORM

22 April 2005

UNESCO Regional Advisor for Cultrue in Asia and the Pascific
UNESCO Bangkok Office
Bangkok,Thailand

Dear Sir:

Zhaoxiang Huang Ancestral Hall, in Foshan (The Project of the conservation of Zhaoxiang Huang Ancestral Hall)

This is to certify that the above building is the keeping property of **the Communal Project Administration**, Foshan, and now it is rent to the Cantonese Opera Museum. I hereby grant permission to **the Academy of Oriental Archi-culture of South China University of Technology** to nominate the above building for the UNESCO Asia-Pacifc Heritage 2004 Awards for Culture Heritage Conservation.

佛山兆祥黄公祠（即兆祥黄公祠古建筑保护工程）。

广东省佛山市公用事业管理局目前为**兆祥黄公祠的**产权保管人，该建筑现租赁给广东粤剧博物馆使用，我据此同意华南理工大学东方建筑文化研究所代表我们申请 2004 年联合国教科文组织亚太区文化遗产保护奖

Sincerely yours,

Du Daquan

Urban Construction Bureau, Foshan (Property trustee)

Renmin Road, Foshan , Guangdong , P.R.China

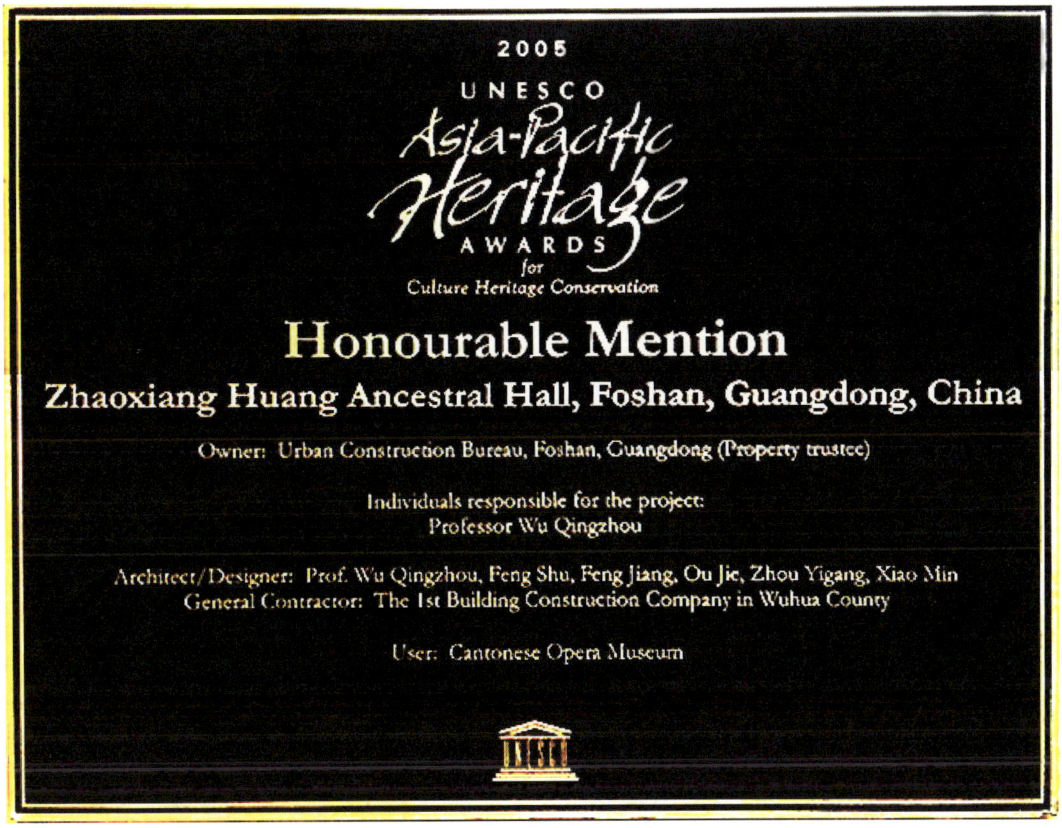

联合国教科文组织

2005年亚太文化遗产保护奖

兆祥黄公祠的保护为佛山一座与中成药发展历程有关、具有历史意义的建筑赋予了新的生命，扎实的历史研究造就了精细的修复途径，用最低限度的改造，采用传统的匠作技艺和材料，确保所有干预措施的可逆性，突出了建筑的原真性。从宗族公共空间到粤剧博物馆的转变在更大范围的社区发挥了无形遗产的价值，为项目贡献了额外的附加价值，同时保障了其在市民社会的长期可持续性。

广州日报
珠三角
PEARL RIVER DELTA NEWS
2006年5月9日 星期二
责任编辑：黄开鹏 Email:gzrbzsj@gzdaily.com 美术编辑：杨泳海
A21

『亚太地区文化遗产保护奖』

兆祥黄公祠昨喜领

图／记者龙成通
文／记者黄慧 通讯员孔耀基

□ 其见证佛山古建筑艺术和当地中成药发展史
□ 为广东省第二个获此荣誉的文物单位

昨日上午，佛山市禅城区兆祥黄公祠内外一片热闹。联合国教科文组织亚太地区文化遗产保护项目协调员来到此祠，参加兆祥黄公祠荣获“联合国教科文组织亚太地区文化遗产保护奖”的颁奖仪式。

据介绍，该祠是继从化广裕祠之后我省第二个荣获此奖的文物单位。

兆祥黄公祠为广东省第二个获此荣誉的文物单位。

兆祥黄公祠见证了当地中成药发展史。

历经80多年风雨沧桑，黄公祠经修复后风采依旧，其占地约3000平方米，建筑平面布局为三进院落，为广府地区的典型祠堂形制。祠堂主体建筑除拜亭为歇山顶外均为单檐硬山顶，面宽、进深各三间，最后一进建筑为两层高，建筑气魄宏大，设计精巧，装饰华美。

见证当地中成药发展史

据了解，兆祥黄公祠历史久远，同时见证了佛山古建筑艺术和当地中成药产业的发展史，而且，其受到当地政府和市民的很好保护和利用，成为粤剧博物馆，呈现更大的文化遗产价值，这就是其获此荣誉的关键所在。

昨日上午，在颁奖之前，联合国教科文组织亚太地区文化遗产保护项目协调员里克·彭先生一行参观了黄公祠和博物馆里的粤剧文物，他们都被古祠的建筑和风格以及其修复后的不变的古貌所吸引，还有几千件的粤剧文物也让他们大开眼界。

据介绍，2003年，修复后的黄公祠被利用为粤剧博物馆，如今馆内分史、艺、人三大部分共13个展区，展出明代、清代至当代的粤剧剧本、木鱼书、海报、戏桥、戏服、乐器，还有早期的粤剧电影、唱片名伶书画等珍贵文物3000余件，展现了粤剧文化丰富的内涵和独有的魅力。

巧用为粤剧博物馆

里克·彭先生是第一次来佛山，也是第一次亲临黄公祠，他觉得现实的黄公祠要比资料上的美，那种古朴的东方建筑以及充满生机的文化气息，让他感到黄公祠获此荣誉是实至名归。

另外，他对黄公祠的修复技术也很赞赏。他说，古祠的修复完全符合评审的要求。

华南理工大学教授吴庆洲表示，在修复黄公祠的时候，经过很精细的测绘工作，作出很详细的方案，如添加构件种类和不可移动的设施，各种技术、安全措施均不改变文物的原貌，清理、拆除后人不合理的添建、改建等，保持原建筑的风格，修补复原。

资料链接

兆祥黄公祠是著名中成药“黄祥华如意油”创始人黄大年的祠堂。从1905年起，历时15年建成，占地面积约3000平方米。它是佛山城区内现存规模最大、形制最完整和装饰最丰富的祠堂建筑群，在佛山市祠堂建筑中极具代表性。

1999年，佛山市政府在黄公祠周围开辟休闲公园，四周已完成绿化，在祠堂前面建起了池塘和休闲广场。同时，佛山市政府筹资修缮黄公祠，重现当年“大宅门”风采，并在此展示明清年代佛山中成药产业辉煌历史，免费向市民开放。

2002年7月广东省人民政府公布其为省文物保护单位。

资料链接

“联合国教科文组织亚太地区文化遗产保护奖”设立于2000年，目的是奖励私人、私人机构及公私合作组织在保护地方遗产、彰显其文化价值方面所作的贡献，奖项分为“卓越奖”、“杰出奖”、“优秀奖”、“荣誉奖”。该奖项每年不定期举行，参加竞赛的项目为古镇建筑、本土建筑、东西混合型建筑等多种类型的古建筑，其条件必须有50年以上历史，必须为私有产权或为私人所租赁，于最近10年内完成，至少经过一年使用的保护或修复项目才有资格申报。

2005年度，联合国教科文组织亚太区文化遗产委员会共收到来自11个国家和地区的34个项目申请，此次有8个国家和地区的14个项目获奖，中国占6个，其中兆祥黄公祠项目获得文化遗产保护奖的荣誉奖。

领奖。

文物点评

在颁奖现场，里克·彭先生宣读有关评判委员会的评语：

兆祥黄公祠是与中医有着历史联系的佛山历史上一个重要的建筑，它的保护工程赋予了这一建筑新的生命。扎实的历史研究促成了这一精心的修复方法：以最低限度的改造，用传统的手工技能和原料，成功转化了所有的干扰因素以突出原建筑的真实性。把该建筑从一个宗族的场所转变为粤剧博物馆，给一个更大的社区带来了无形的文化遗产价值，带给这项工程更多的价值，同时也确保了它在市一级的领域中长时期的稳定地位。

香港国际中医药总会永久会长、黄大年曾孙黄凝鎏：

黄公祠本是祖父为纪念曾祖父所建。它本是为弘扬中医药文化，如今它不但得到最好的保护，还被有效地利用成为广东省粤剧博物馆，成为宣扬粤剧文化的一个重要地方。

家祠有如此光华的今天，让我很感动，希望它将为更大范围的社区服务，宣传更多的传统文化。

广东粤剧博物馆挂牌仪式

四、相关研究论文

佛山兆祥黄公祠的地方性材料、构造及修缮举措 [1]

Domestic Materials, Tectonics and Some Measures in the Conservation Project of Zhaoxiang Huang Ancestral Hall in Foshan City

冯江，郑莉

Feng Jiang, Zheng Li

1. 本论文原发表于《南方建筑》2008 年 5 期，本次有所修改。

【摘要】兆祥黄公祠修缮工程获得了联合国教科文组织2005年度亚太文化遗产保护奖荣誉奖，该祠堂是佛山老城内所保存的规模最大的祠堂建筑，其材料、构造和细节做法在广府地区清末民初的同类建筑中较为典型，本文针对该建筑中的一些地方性的材料和工艺进行了较为详尽的探讨。在80余年的使用过程中，兆祥黄公祠受到了较为严重的破坏，本文基于对古建筑的修缮原则与修缮技术之间关系的理解，回顾了整个修缮工程的过程，就其原有做法和修复中采用的做法进行了比较。

【关键词】兆祥黄公祠　修缮　地方性　构造　材料

【Abstract】Zhaoxiang Huang Ancestral Hall is the largest reserved ancestral hall in the national historic city Foshan. The material, tectonic and details of this building is very typical among those ancestral halls built in the beginning of 1900s in Guangfu Area. The conservation project of Zhaoxiang Huang Ancestral Hall won the 2005 Asian-Pacific Cultural Heritage Award，UNESCO. The article makes a review of the whole restoration process based on the comprehension of relationship of conservation techniques and principles, and discusses some domestic materials, crafts and tectonic traditions.

【Keywords】Zhaoxiang Huang Ancestral Hall, Restoration, Locality, Tectonics, Material

2. 联合国教科文组织颁奖词中对该修复工程的评价。

兆祥黄公祠修复工程是2005年度联合国教科文组织亚太文化遗产保护奖的获奖项目，研究者在修复设计和施工过程中对地方性建筑材料、工艺给予了较多的关注，以详细记录和深入研究为基础，“用最低限度的改造，采用传统的匠作技艺和材料，确保所有干预措施的可逆性，突出了建筑的原真性”[2]，是一次恪守本分，以合理的花费进行修缮，不进行过度粉饰而又凸显建筑地方性，重新赋予文物建筑活力的尝试。

一、兆祥黄公祠及其修缮工程概况

佛山是岭南中成药的发祥地之一，老城区内有很多与之相关的历史遗迹，兆祥黄公祠即为其一。该祠是著名中成药“黄祥华如意油”的创始人黄大年先生的祠堂，也是佛山老城现存规模最大、形制最完整的祠堂建筑。

黄大年，字兆祥，原经营“五福”灯饰店，后改营药油，祖铺设于祖庙大街文明里，始创于清咸丰年间，黄大年四子黄奕南根据祖传药方，制成既保存原有药方特效又能医治一般常见疾病的药油，可以搽食兼用，特别是对四时感冒、肠胃不适、风痰咳嗽以及小儿腹痛、烫火刀伤、蚊虫蜇伤等疗效甚佳，定名为“黄祥华万应如意油”。据说，光绪十年（1884年），李鸿章巡视广东，随行眷属因水土不服而染疾，服搽如意油后病除，李鸿章赠给黄奕南“韩康遗业”四字[3]，黄祥华如意油从此驰名，在广州、汕头、江门、上海、香港、新加坡等城市纷纷开设店铺，黄祥华成为佛山首富。据黄奕南之孙黄凝鎏回忆，这座祠堂是其祖父为曾祖父所建[4]，也就是说，是黄奕南为自己的父亲黄祥华修建了这座家祠。

3. 此故事见于20世纪30年代的“安宫牛黄丸”说明书，引自《佛山日报》2006年4月21日。

4. 据《广州日报》2006年5月9日。

该祠堂位于佛山老城的折线形城市空间发展轴上，在现福宁路与建新东路交叉口的东南方向，始建于清光绪三十一年（1905年），建成于民国9年（1920年）。祠堂坐西朝东，占地约3060m^2，建筑部分东西长67.57m，南北宽31.40m。建筑平面布局为三路四进，中轴线上布置头门、拜亭、正厅（祭堂）、寝堂、后座等主体建筑，较为独特的有2点：一是在一进院落与享堂之间有一座歇山顶的拜亭，二是在寝堂之后尚有一进天井院和一座两层的主体建筑，且后座建筑从细节做法和材料上明显不同于前几进建筑，建设年代应稍晚。两侧的边路对称布置衬祠，以青云巷与主体建筑连接，为广府地区清末民初祠堂较为典型的做法。中路上的主体建筑除拜亭为歇山顶，最后一进2层楼外均为单檐硬山顶，面宽、进深各三间。建筑气魄较宏大，设计精巧，装饰华美，在南番顺地区的祠堂建筑中颇具代表性（图1）。

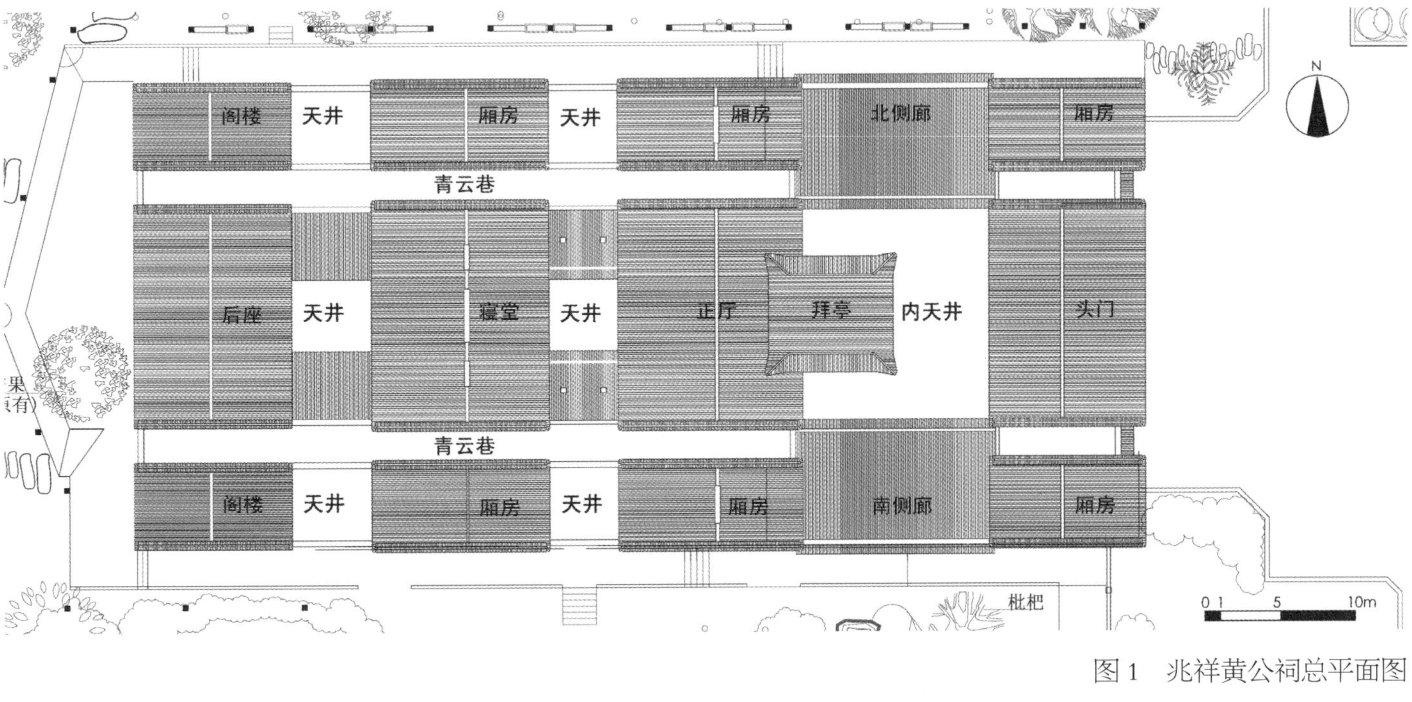

图1 兆祥黄公祠总平面图

抗战期间，各地的黄祥华如意油店铺纷纷倒闭，后店铺改号迁往香港。[5]在80余年的历史变迁之中，黄氏家族的后人大多移居港澳和海外，兆祥黄公祠亦失去了原有功能，长期被闲置，曾先后作为国民党南海县党部和军队的驻地，两度进行了不合理的加建。“文化大革命”中，祠堂的很多构件遭到严重破坏，头门檐下梁架上的木雕遭到砍斫和火烧，大量装饰构件失散。1985年，一名部队战士在学习开汽车时撞折了头门的石檐柱，导致头门屋顶大面积坍塌。但在经历了战乱、社会变革和大规模城市建设活动的冲击之后，兆祥黄公祠仍能幸存并且基本保存完整。

5. 黄氏后人继承了祖业，如意油的制炼传承是：黄元吉—黄兆祥（黄祥华）—黄奕南—黄颂陶—黄凝鎏—黄启昌。1950年，黄凝鎏在香港分店设厂生产，供货给港澳各药房销售，并以“黄祥华流行堂帆船牌万应如意油”的名称，先后在香港、新加坡、马来西亚、印度尼西亚、台湾等国家和地区注册，从此“黄祥华如意油”又行销世界各地。黄凝鎏现为香港药行商会会长、香港国际中医药总会永久会长。

1989年，兆祥黄公祠被公布为佛山市级文物保护单位。1992年，祠堂要被驻军作为危房拆除，经过多方努力，部队最终迁出，祠堂得到了保护。1999年，佛山市政府在兆祥黄公祠周围开辟兆祥公园，并多方筹集资金，支持修复该祠堂。修缮工作于2000年正式启动，建立了产权托管人、未来使用者、研究机构、古建专业修缮机构、文化热心人士、政府部门共同参加的多方会议机制，并委托吴庆洲教授主持，由华南理工大学东方建筑文化研究所承担测绘和修复设计。本次修复探索了如何利用有限的资金，区别于本地区古建保护中常见的粉饰方式，采用恰当的手段对经历了多次加建、改建的文物进行更有效的保护和维修，重新强调了传统材料、工艺和传统建筑类型的价值，也为社区的公共生活作出空间和文化上的贡献。整个修缮共历时2年，于2002年顺利完成，后交付广东粤剧博物馆使用，成为一座地方戏曲博物馆且运行良好。

二、兆祥黄公祠中的一些地方性材料

1. 木

清末民初，珠江三角洲地区出现了一次祠堂建设的高潮，而且与以往主要集中在乡间不同，城市之中也出现了较多新建的祠堂（包括以书院之名出现的合族祠），大量优质木材被用于建造祠堂等建筑。需要说明的是，多种木材并非产自本地，但在广府地区的重要建筑之中使用较多，亦成为颇具地方色彩的建筑材料。因为广府地区潮湿的天气易使木材糟朽、腐烂，而且白蚁向来是木结构的大敌，所以产自东南亚的略带酸性的致密木材就受到了广泛的欢迎，其中坤甸木、东京木、铁力木等最为常见。在兆祥黄公祠中，所采用的硬木主要有坤甸木、东京木、波罗格和花梨木。

坤甸木，拉丁名*Eusideroxylon zwageri*，英文名Borneo ironwood，俗称铁木、铁檀，得名于印度尼西亚加里曼丹岛的西部城市坤甸。坤甸是华侨的集中地，广东从坤甸进口的这类木材较多，人们就称之为坤甸木。坤甸木心材色泽黄褐至棕褐色，边材金黄色，心材与边材区别明显，心材强度甚高，气干密度约1.19g/cm^{3}左右。坤甸木结构细且均匀，纹理多斜行，材质硬重，强韧耐腐，抗蛀力强，为有名的高强度结构用材，其优点是木材坚硬，油漆后光亮性好，且不怕潮湿，置于潮湿处不会腐蚀，而浸于水中则越坚实，是重要建筑中常采用的木材，清代广州的船舶用材也多采用坤甸木。[6]

6. 蔡易安编著：《清代广式家具》，上海：上海书店出版社，2000。

东京木，或称格木，拉丁名*Erythrophleum fordii*，树高可达25m，出产于非洲、东印度、泰国、越南、苏门答腊一带。格木是较早进口的名贵木材之一，过去大都从越南进口，越南北部红河三角洲地区旧称东京（Tokin），广东遂称此种木材为东京木。东京木有深棕色和棕灰色的条纹，心材红褐或略带黄，边材黄褐色，木材纹理交错，有光泽，结构均匀，质地坚实，较难切削，刨光和油漆后光亮性良好，能抗蛀耐久。历史上东京木进口数量较多，因其树材高大，材质坚韧，具有较强"担力"（抗弯能力）和"顶力"（抗压能力），广东许多大型建筑和园林建筑常用之作梁柱、大门板等，其他如舵杆、秤杆、乐器等也常用东京木加工做成，虽历百数十年仍坚实如故。[7]

有人认为东京木古称为铁力木，但此说有争议。铁力木亦称铁木（Iron wood），主要产于印度尼西亚东部、锡兰、缅甸、斯里兰卡、马来西亚和印度东部，我国云南西双版纳和广西的容县、藤县也有出产。明曹昭《格古要论》中说："铁力木，出广东。色紫黑，性坚硬而沉重。东莞人多以作屋。"铁力木杆长而大，心材红褐色，材质坚硬，常用于家具、建筑和造船业，屈大均在《广东新语》"木语"中提到："铁力木，理甚坚致……广人以作梁柱及屏障。"肇庆梅庵和顺德碧江尊明祠的主要用材均为铁力木。

7. 蔡易安编著. 清代广式家具[M]. 上海：上海书店出版社，2000。

波罗格，拉丁名*Intsia bijuga*，被称为金黄色泽系美术工艺木材。波罗格结构较粗而质坚，纹理为参差短直沟纹木材，油漆光亮好。由于木材重而硬，强度高，耐腐性强，并且具有一定的花纹，所以多用于要求木材耐久、强度大和有装饰性的建筑构件、高级家具、细木工、地板等。

花梨木，又称花榈，拉丁文名*Dalbergia odorifera*，其中明代常用的为"黄花梨"，广州称为"降香"，清代中叶以后因黄花梨材小源缺而被新花梨取代。宋赵汝适《诸蕃志》中即有记载："降香出三佛齐（苏门答腊）、阇婆（爪哇）、广东西诸郡亦有之……"清代中叶以后大量花梨木材输入广州等地，且进口的多为重达数吨的巨材，主要用于制作硬木家具、乐器和工艺品，亦用于建筑装饰。

另外，广府建筑之中带雕刻的装饰性木构件常用柚木（拉丁名*Tectona grandis* Linn），因其纹理均匀，直而清晰，使用时尺寸稳定，油色之后呈现出自然美。

在兆祥黄公祠中，坤甸木主要用于柱、大梁、斗栱、瓜柱、驼峰等重要结构构件，而东京木被用来加工为檩子和桷板。头门梁架为三间十三架，拜亭为卷棚歇山顶。正厅、寝堂和后座二层结构均为三进柱网，檐下均为六檩卷棚，中跨为九架，脊檩至后檐墙间为九架。部分主要木结构构件的尺寸见表1。

兆祥黄公祠部分木结构构件尺寸一览 表1

部位	心间脊檩		檐檩	大梁		后金柱	
	跨度（m）	直径（mm）	直径（mm）	跨度（m）	高（mm）	高度（m）	底径（mm）
头门	5.76	380	240	4.42	395	6.86	443
拜亭	–	–	240	6.85	335	–	–
正厅	5.76	390	240	6.36	410	7.62	448
寝堂	5.76	390	240	6.02	410	8.07	400
后座	5.76	372	236	5.10	392	5.41/4.91	440/440
厢房	5.53	270	180	–	–	–	–

注：1. 拜亭为卷棚歇山顶，四角均为石柱。
2. 后座后金柱尺寸分别表示一层、二层的尺寸，梁为二楼的尺寸。
3. 头门为纵墙与后金柱之间的六架梁，正厅、寝堂和后座主楼的梁均为前、后金柱之间的九架梁，其长度值包括前金柱半径和出后金柱的梁头（图2）。

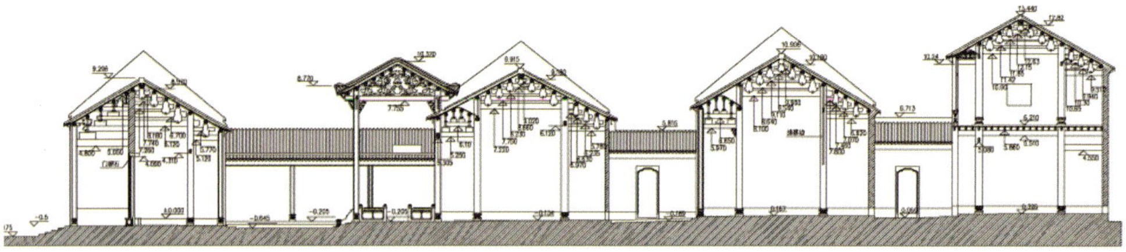

图 2　兆祥黄公祠心间纵剖面（左侧为头门）

图 3　头门封檐板局部（冯江摄）

整个祠堂共有木柱22根，均用坤甸木，各主要建筑后金柱尺寸见表1。

除坤甸和东京木之外，祠堂中还使用了波罗格、柚木、花梨木、杉木等木材。波罗格用于雀替等兼具装饰性和结构作用的构件，柚木则用于需要精细雕刻的构件，例如挂落、封檐板等。阁楼的楼板和承重的木枋采用的是杉木，两块企口的木楼板中间夹着竹筋，木扶手和栏杆均为杉木。祠堂的大门的用材是花梨木，其尺寸为121cm×414cm；其他各处木门均已缺失（图3）。

2. 砖

兆祥黄公祠墙身所用均为青砖，用于墙面砌筑和装饰，规格有多种。

主要用于头门正面磨砖对缝的青砖尺寸一般为280mm×110mm×（62/40）mm，剖面形状为不规则的楔形，除正面磨得平整光洁之外，其他各面均凹凸不平；头门山墙东端转角处的水磨青砖平面为梯形，从坍塌之后的断垣可以清楚地看到此处青砖的形状和构造做法（图4、图5）。

现场记录下来的用于砌筑承重空斗墙的砖有5种规格（单位均为mm）：

① 253×110×57；② 253×110×53；③ 254×110×57；④ 257×110×61；⑤ 262×110×62。

这几种规格比较接近，有可能只是同一种规格在烧制时产生误差的结果，其尺寸比东莞大青砖小。

青砖表面颜色较深，看起来似乎不利于反射阳光，而白色是常识中比较适合炎热气候的，正如20世纪80年代岭南派建筑所作的选择。但实际上这是没有认真考察当时建筑语境的一种误判，在街巷很狭窄的情况下，黝色的青砖吸收了热量，而没有反射给行走在巷道上的路人，墙身上的热量也不会影响室内，因为有眠空斗墙中空气层的存在，外墙所吸收的热量对室内的影响并不大。

另外，墀头及边路衬祠天井中的墙壁上有多处砖雕。

图 4　头门局部坍塌之处所见的墙身断面（冯江摄）

图 5　头门墀头砖雕（冯树摄）

3. 方砖

现有的室内地面铺砖共3种：灰色尺二方砖（360mm × 360mm × 6mm）、红色尺四大阶砖（470mm × 470mm × 10mm），以及200mm × 200mm × 5mm有几何装饰图案的陶瓷花砖（图6）。

图 6　拜亭地面花砖图案一角
所示尺寸：400mm × 400mm

兆祥黄公祠主要朝向为东，根据习惯，次要房间、各阁楼处为尺二方砖斜45°错缝密铺，不勾缝；中路主要单体建筑采用红色尺四方砖错缝拼缝密铺。因为宽度和厚度合适，红色尺四方砖也被用来作围墙叠涩檐口的压顶，以及陶瓶栏杆的压顶。

拜亭地面采用陶瓷花砖，装饰性很强。

4. 瓦

兆祥黄公祠的瓦件一共包括仰瓦（板瓦）、筒瓦、瓦当和滴水4种，全部为陶制，滴水和瓦当的端面为釉面，滴水及瓦当的后部、仰瓦、筒瓦均为陶瓦本色。

板瓦尺寸为210mm × 210mm × 10mm，颜色为淡赭色。筒瓦的长度为210mm，宽度（直径）为110mm。滴水和瓦当的端面分别为如意形和圆形（图7）。

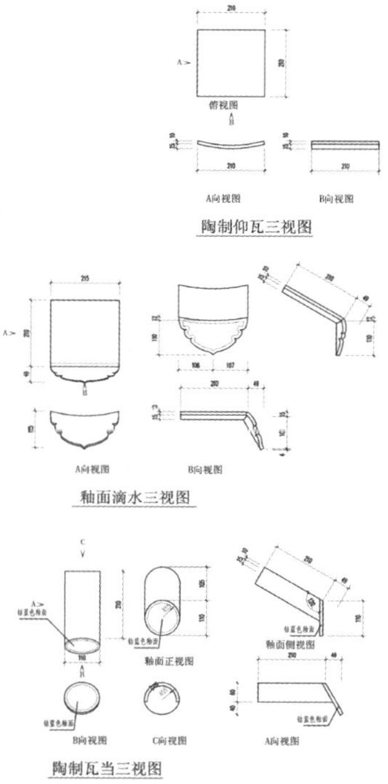

图 7　瓦件一组

5. 石

岭南多雨，而石材耐潮湿，所以在许多容易遭受雨水的部位多采用石构件，例如檐下的石柱、柱础、勒脚。为避免水分通过毛细作用导致木门窗变形，门边和窗户四周都用石头包边。石材的热物理性能对于地面降温有利，故庭院、天井、青云巷等处的室外地面都采用石材铺砌。黄公祠的石构件主要包括各进基座台明、塾台、石勒脚、檐柱、全部柱础、石驼峰、虾弓梁、墙体的勒脚、门边包石、窗边包石、石门臼、水井四周、露明地面的铺地条石、石栏杆、集水口以及各处的石雕。

中路上各进建筑的檐下均用石柱支撑，其中最长的是支撑拜亭的4根花岗石檐柱，长度达到了8.21m，前檐柱截面尺寸为300mm × 300mm，而靠后的2根与正厅共用的石柱截面尺寸为400mm × 400mm。

铺地条石均为花岗石，规格统一为350mm宽，长度则变化较大，最长的条石为拜亭前石阶的整块踏步，长达5.6m（图8）。

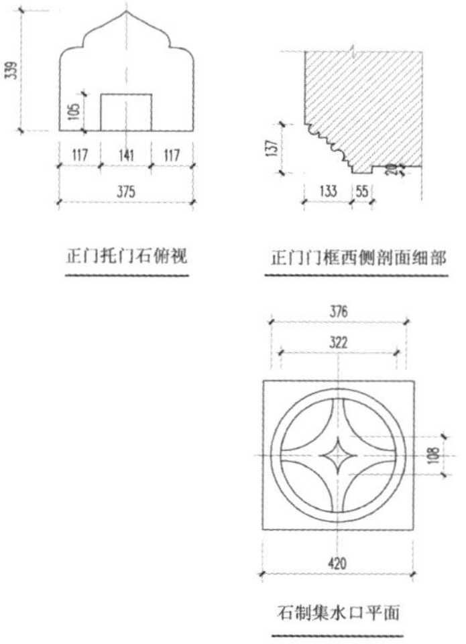

图 8　石构件一组

6. 铁

铁件主要见于窗户的防护栏杆、小五金件和排水天沟、水落管，栏杆多用直径30mm的圆铁，至今只有一层很薄的铁锈，局部还可看到柔和的光泽。

三、兆祥黄公祠中的一些地方性构造

地方性的工艺和建筑材料、建筑形制密切相关，而材料、形制都受到气候的影响。同时，制约建造的重要因素还包括地区的经济发展、技术状况和文化、审美取向等。

兆祥黄公祠的头门采用了类似一门两塾的形制。古老的门塾制度虽然在明代的广府祠堂中就已出现，但到清代才成为广府地区的重要祠堂中较普遍的做法。塾台的正面通常会采用石雕装饰，成为重要的展示内容，而檐下用石柱则是对本地潮湿多雨气候的一种适应，头门的前檐石柱立在塾台之上。兆祥黄公祠的头门是一种折中的做法，可以看作是塾台和基座的结合，看似双塾，但因为心间台阶的顶部、心间地面的标高与次间标高相同，所以实际上仍然只是基座。

拜亭也是适应湿热气候的结果，民间也有将其称为接旨亭或者仪亭的，顺德乐从大墩村的梁氏家庙、花都朗头村的友兰公祠都有拜亭。此处拜亭屋顶虽为歇山形，结构上却用卷棚的做法，同样的情形也见于顺德西山庙的覆龟顶。

为形成良好的微气候，广府地区的民间建筑群中广泛使用青云巷和小天井的做法，屋面则有风兜等构造发明。因为阳光强烈、空气温度高，所以如何利用建筑来形成阴影进而通过空气温度差形成微风，以及利用建筑材料和建筑构造来帮助降低空气中的温度就成为本地区建筑需要面对的问题。

因为常年湿度高，降雨频繁，所以在构造上如何隔绝湿气，阻止毛细作用对木门窗、地面和墙身的破坏也导致了一些具有本地特点的构造做法的产生。

1. 砖作

由于祠堂有着重要的精神意义，所以建筑正面的装饰非常考究。兆祥黄公祠三路建筑正面的所有墙面均采用磨砖对缝的工艺，以全顺的方式砌筑，以获得光洁、整齐、纯粹的视觉效果，且正面不开窗。

水磨青砖是专门用于装饰的一层构造，实际上边路厢房的东侧外墙达到了400mm，头门的外墙厚度更是达到了470mm。头门外墙的构造共4层，中间2层为承重的空斗墙，所用青砖的规格与用于饰面的水磨青砖不同，为254mm×110mm×57mm左右。而磨砖对缝的砖也包括了5种规格和2种做法，比较特殊的是转角处采用了俗称“碰角”的做法，即2个相互垂直的墙面上的砖以梯形的尖锐角部相接，而非一般情况下采用的丁砖、顺砖相互搭接的做法。

其他外墙一般厚度为400mm，采用5层顺砖、1层丁砖的做法，而室内则是全顺的淌白做法，表面并不批荡。[8]这也是广府地区祠堂外墙的通常做法，外墙的5顺1丁砌筑方式形成的有眠空斗墙既比较经济，又能很好地适应本地的湿热气候，而内侧的全顺砌法则主要是为了更好地承重。

8. 广东称墙面粉刷为批荡。

室内地面的大阶砖下，是厚达1m的沙，没有土、石灰等任何其他材料，这种做法有利于隔绝来自地基的湿气。

2. 大木作

兆祥黄公祠的木结构体系是广府地区清末民初祠堂建筑中较为常见的一种，正厅、寝堂前檐下的一进均为卷棚，卷棚后檩直接落在前金柱上；后两进为抬梁式的做法，横向（此处为与朝向平行、与屋顶坡向垂直的南北方向）用斗栱，而纵向梁架则用瓜柱，但各主体建筑中出头门之外并未使用水束的做法。

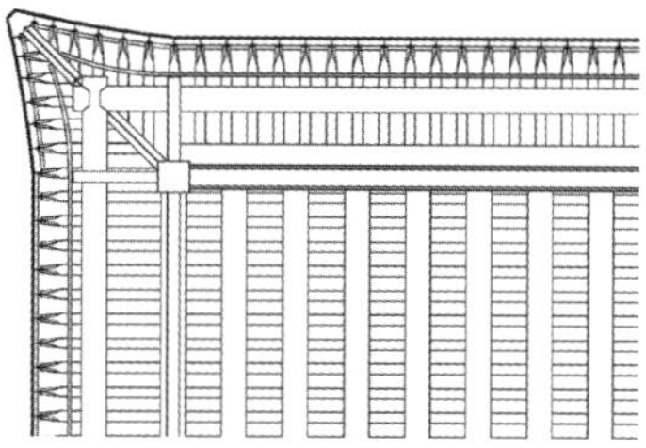
图 9　拜亭角部仰视图

较为特别的是拜亭和后座2层高的主楼。拜亭为卷棚歇山顶，屋顶直接用桷板承托瓦面，筒瓦位置与桷板对应，而檐口则用桷头来完成造型和承重，和仔角梁一起使翼角形成很细微的起翘（图9）。后座的2层使用了各自独立的木梁架体系，一楼用方形的密集木肋支撑楼板，二楼的金柱直接落在楼板上，类似官式楼阁建筑中的缠柱造。

3. 石作

石材在广府地区的建筑中发挥着重要的结构和构造作用，产生了许多成熟的做法。

兆祥黄公祠是浅基础，所以地脚梁就是带槽的石条。而所有墙身底部都有石勒脚，从正面来看，边路的勒脚和基座的高度达到了1705mm，中路则有760mm高的基座，基座的侧面和顶面均为石板，塾台上的石勒脚高度为1125mm。头门的前后檐柱均为石柱，心间无阑额而只有木雕封檐板，如此则获得了开阔的视觉效果，可以清楚地看到祠堂的大门。次间则有虾公梁，应是由月梁演变而来，虾公梁架在檐柱和山墙之间，梁上用石狮托住檐檩。因为青砖的毛细作用较为明显，靠近地面的位置容易受潮，而且砖墙受潮之后容易结硝和长苔藓，影响承重，所以墙身底部使用石材可以有效保护墙身。

黄公祠门窗四周均有石边框，一般门窗的石框宽度均为110mm。这些包边石和石勒脚有着相似的作用，顶部的石条兼有过梁的作用。如果不设这些石框，则砖中的细孔会将水分渗透到门窗的木框上，导致木门窗的变形。祠堂大门的门边石侧面宽度为1.07m（含石门框），顶部宽度为1.46m，除了帮助固定巨大的门扇之外，亦有炫耀财富之意。

石柱础则可以有效保护木柱，兆祥黄公祠共有仰莲、圆锥和雕刻有杨桃的鼓镜形3种，均为圆形且较敦实，而石柱下的柱础则为方形，较为纤秀（图10）。

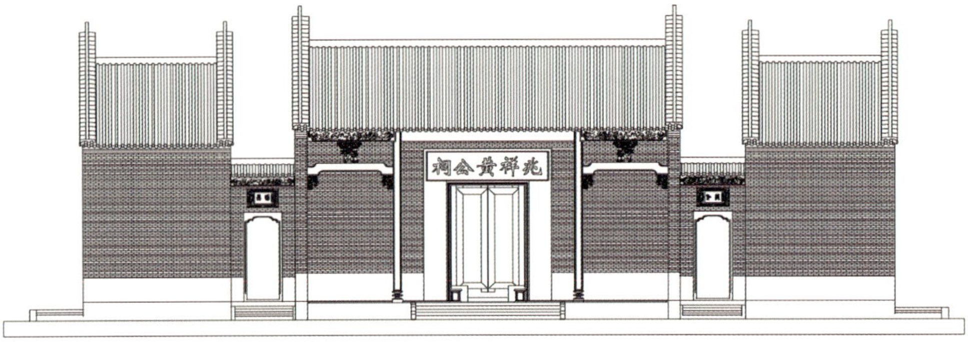

图 10　兆祥黄公祠正立面修缮设计图

图 11　二进北厢房上的风兜及其剖面（冯江摄）

图 12　头门檐口断面（冯江摄）

4. 瓦作

广府民间建筑的屋面做法较为简明，板瓦直接放在桷板上，是朝室内露明的，因此要求瓦面整齐、美观，施工仔细。瓦从心间起铺，中轴线上铺瓦坑，施工误差在两尽端分配。兆祥黄公祠心间共有29个瓦坑，次间15个，瓦垄的筒瓦上糊黑色裹垄灰。

兆祥黄公祠的屋面多处使用了卷棚，除拜亭和南北分坡的侧廊之外，在二进厢房还采用了风兜的特殊构造（图11）。风兜是通过2层不同标高的屋面来引导顺坡而起的凉风进入室内的做法，是增加自然通风的有效方式。

而黄公祠的檐口所用仰瓦有7层之多（图12），搭接方式较为复杂、巧妙，对垂铺时的顺序、屋面的平整度和瓦垄的直顺度都有较高的要求。各硬山脊顶亦用瓦檐，形成很小的保护山墙的排山勾滴。

5. 雕饰与油漆

兆祥黄公祠头门前檐下的梁架和各处的雀替、斗栱、驼墩都有丰富的雕饰，其余部分则较为简洁。石雕、砖雕和灰塑虽均有采用，但使用较为节制。祠堂的色彩整体上较为肃穆，梁架为深褐色。

四、兆祥黄公祠的破损情况及修缮举措

1. 破损情况

1）总体格局

整个祠堂共4进，头门遭到严重破坏，除头进主体建筑后部被砖墙封堵，二进主体建筑后墙被多开了2个洞口之外，各主体建筑基本保存原有格局，但2个边路被改建较多。多个小院、天井被覆盖，通向户外的门被封堵，且存在大量的加开门窗洞口的现象。

2）屋面

因人为破坏和长时间未检修，屋面普遍存在破损、糟朽、漏雨等现象。从现状看，屋面的各种损坏类型包括：

（1）头门处屋顶大面积坍塌。

（2）屋顶被改建。头进南侧厢斜屋面和卷棚交接处屋面被改建，改建处存在多处不合理的构造，使用了与原来不同的材料和做法，且手工粗糙 。巷道上有多处加建的小屋顶。

（3）多处屋顶存在漏雨现象，比较集中的是头门两侧厢、拜亭、正厅。

（4）桷头、桷板糟朽。因为经雨水浸泡较多且易受风的影响，屋面下部的瓦垄灰脊、瓦面普

遍已经松动，桷头普遍存在糟朽现象，已经成为安全隐患。

3）墙身

墙体的主要破损情况：

（1）头门处墙身大面积倒塌，成为断垣。

（2）墙身变形，较明显的有3处，头进北厢房东侧墙身顶端有宽约10cm的裂缝，南侧墙身有明显倾斜，三进厢房墙身中央部分向外突出。

（3）部队使用期间多处添建了青砖、红砖、木板隔墙，且多开了大量门窗洞口，添建了大量门楣和窗楣；朝外的各天井侧门被用砖堵塞。

（4）大部分墙面被加以石灰批荡。

4）木结构构件

木柱现状基本良好，但底部普遍有裂缝，有多根遭到不同程度的白蚁蛀蚀，其中有2根相对严重。石柱除头门处被撞断的一根之外，其余基本完好，但柱础角部细小的崩缺较多。

除头门梁架被严重破坏外，梁架保存基本完好。局部遭受蚁害，并有部分被改建过，局部有数根增加的檩。

5）装饰构件

大部分木装饰构件已经不存，主要在“文革”时期失散，现已流入民间收藏家之手，尚存的构件中，头门下的金木雕在“文革”时遭到红卫兵的刀削斧斫，而雀替、斗栱、驼墩受到的破坏并不大。

屋顶装饰被破坏，所有建筑正脊均被破坏，但无历史照片和图纸记录，有人认为应该是石湾陶塑，有人认为是龙船脊，无法确知其旧貌。头进院落两侧建筑的压檐仅有局部残留。后座屋顶原有装饰构件被破坏。

2. 修复原则

早在2000年3月，参与修缮的各方就确定了“所有修复措施都必须基于对建筑的考古学和历史学研究，基于对历史原真性的保护”的首要原则，因此，不添加构件种类和不可移动的设施，各种技术、安全措施均不改变文物的原貌。在此基础上，派生了其他的原则：

（1）经审慎研究后清理、拆除后人不合理的添建、改建部分，保持原建筑的风格。本原则适用于部队使用期间加建的门窗、墙、屋顶、阁楼等。这些加建部分破坏了祠堂的本来建筑面貌，损害了祠堂的整体性和艺术价值、历史价值。因为加建、改建之处较多，在审慎地进行辨别、研究，根据材料、工艺、装饰风格、残留的细小痕迹等确定被改动的部位之后，拆除多余的不合理的添建、改建内容。

（2）关于建筑材料的原则：对明显可考证且易于修复、恢复的部分或者影响建筑物正常使用、意义重要的部分进行整修、恢复，尽量保持原有构件与材料，修缮如故；如必须更换则采用相同材料，原构件无法再用的补以新构件，修补复原，并与旧材料进行区分。

（3）无法考证且不影响建筑物正常使用的缺失构件不进行复建。部分装饰构件已经全部佚失，虽然其尺寸、大致轮廓可以根据残留的痕迹和同时期本地祠堂建筑装饰的一般风格大概推断，但其真实面貌已经无法考证，因为不影响建筑物的正常使用，依照本原则，不予修复。例如，现已全部被破坏的正脊脊饰不再补回，但对于其残留的痕迹和构件则予以仔细保留。由于修复之后兆祥黄公祠将作为博物馆使用，而不再具有宗祠的功能，故对于陈设和不可考的装饰不再恢复。

（4）排危和延长建筑物寿命的原则：消除或尽量减少各种安全隐患，在尽量不影响外观的情况下，对祠堂的结构、排水系统、防火措施、防雷等技术内容进行全方位的检查和改进，尽可能避免灾害的发生，在万一发生灾害时使其保持一定的应变和抵抗能力。所有因为技术改进而增加的构件（例如结构支撑、消防栓等）尽量独立在原建筑之外，并且视觉上可以明显区别于原有建筑。

（5）为了满足现代社区活动需要的空间，设计充分考虑了文物建筑对于工程措施的敏感性，包括空间尺度的历史感以及装饰物、设施的环境感。建筑的使用计划是作为博物馆，在合理安排展品和参观流线的前提下，现状的空间布局可以实现这一新的使用功能，因此本项目强调了对建筑物现状的充分尊重。对于现代使用功能不可避免的新增设施（例如展览、照明、消防等），本项目要求在降低其对于历史环境的视觉冲击性的同时，保留必要的可识别性。强调不在建筑遗产中增加任何新的装饰物，对于为使用目标而增添的少数具有展示价值的装饰物必须是可以移动的，不得固定于建筑物之上。

（6）对所有房间、院落、结构构件、门窗和其他独立构件进行清晰、准确的技术编号，建立细致的技术档案。并采用“技术交底在前，施工在后，样板引路”的施工程序。修缮工程在消除不安全因素，恢复建筑物原真状态的措施中，强调了技术手段的一致性，即采用传统的材料、传统的工艺来完成所有修缮工作，但是留下了可辨认的痕迹以及详尽的记录。

3. 修复中的一些举措

遵照修复原则，整个修复工程中进行了关于地方性材料、构造和施工方式的诸多尝试，既包括对这些材料和工艺的采用，也包括凸显历史面貌的表现力。

祠堂的修缮大量采用了地方性传统工艺，例如磨砖对缝、砖雕、木雕、石雕、灰塑等。修缮过程中合理运用了传统技术，例如地方的传统防潮做法、木作和石作工艺，延请传统建筑匠师，并将其技术传授给有志于历史建筑保护的年轻匠师。

此处仅对本地古建筑修复中常见而又经常不能得到妥善处理的几个问题展开讨论。

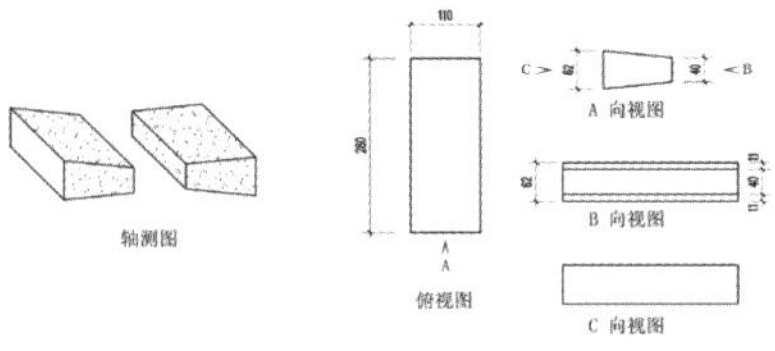

墙面水磨青砖轴测图与三视图

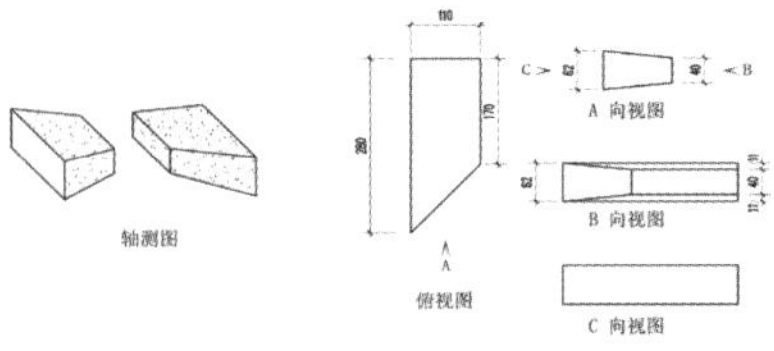

转角处水磨青砖轴测图与三视图

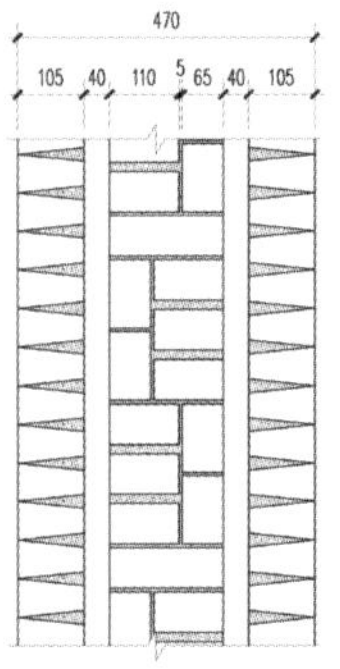

石砌法——两面水磨青砌砖

图 13　水磨青砖墙所用材料规格与头门山墙墙身构造

1）磨砖对缝工艺

磨砖对缝的工艺流程是：切块、修边、粗磨、开槽、细磨。因为整个祠堂一共有5种规格的青砖，加工好的青砖按照规格摆放整齐，施工时“对号入座”，墙面丝缝严密，表面的平整度、垂直度均控制在3mm以内。因为施工方在磨砖环节采用了计时工资而非计件工资，所以工人对青砖的打磨特别仔细，施工之后的墙面效果十分理想（图13）。

2）对墙面后加批荡的处理

在本地的古建筑修复中，使用灰泥对原有清水砖墙进行批荡然后画线是常用的做法，这种做法对于青砖风化较严重的墙身有较好的保护作用，但是对于青砖保存较好的建筑来说，却导致了砖不能正常“呼吸”，而且损坏了墙面的表现力。在兆祥黄公祠，选择了去除所有现有的墙面批荡，而且不再另加批荡。因为各进主体建筑墙面均为青砖磨砖对缝，次要部分为淌白做法，外观效果、物理性能良好，另加批荡反而会破坏建筑的本来风格，所以小心去除现有批荡之后，对墙面进行清洗，还墙面以本来面目。

3）头门梁架

因为头门梁架发挥着重要的结构和精神功能作用，修缮中使用倒塌之后保存下来的原有梁架构件，已经不能再使用的斗栱全部按照原样恢复。封檐板基本被保留，但其上装饰受到不同程度的破坏，补回缺失的封檐板，但对板上已被破坏的装饰不再修复。

对已经断裂、规格为250mm×250mm×6000mm的檐下石柱，重新制作，采用井字木架配合2个手拉葫芦的方法，一次吊装归位成功。

4）脊檩构件吊装

后座主楼脊檩被雨水浸泡受到严重损坏，需要进行更换。因为难以寻找到同等规格的旧坤甸木，故改用樟木代替。后进庭院狭小，难以用现代起重机械进行拆卸和安装，且二楼次间的墙壁均不拆除，3根长约6m的大梁只能用相对灵活、简便的方式进行吊装。施工中采取搭木架结合手拉葫芦的方法，第一步先将脊檩从天井院中吊至棚面，第二步再吊至屋顶榫接安放。

5）装饰

对于“文化大革命”中被斧凿的头门梁架木雕，因破坏严重且可作为建筑所经历的历史遭遇的遗痕，保持被破坏之后的状况，不另外修复，以展示不同历史时期的痕迹。

正脊不再恢复，脊尾饰按照现保留的一个完整的脊尾修复，草龙因为破坏较严重且影响观赏进行修复。灰塑、墀头被破坏处不再修复，而只是进行结构加固和表面清理，将残留的构件安装回原处，但并不进行过度的粉饰，如头进庭院侧廊上的压檐（图14）。

除入口大门外，所有满洲窗、木门均已佚，根据地面和墙面痕迹及保存的石门臼恢复门扇，但装饰不再复原。格栅窗有局部留存，按照残留局部的式样恢复。门窗边的石包边因为有构造上的作用，全部恢复。

木构件的油漆经过反复的现场比对，制作了多个样板，最后选用酯胶磁漆和醇酸磁漆，其中主调采用朱红：黑＝9：7的比例，油漆干后，色彩与原色彩接近。

图14　侧廊屋顶压檐装饰修复前后（冯树摄）

结语

兆祥黄公祠内采用了地方性的材料和工艺解决了原有的结构、构造和功能问题，而且凸显了材料和施工工艺的价值，因为整个修复工程并没有进行过度的装饰和全面的复原，修缮费用比预算大为减少。略显遗憾的是缺少正脊，使得建筑看起来不够完整，但是局部装饰性构件的残缺反而受到了参观者和博物馆的好评。

修复完成之后，兆祥黄公祠被移交给广东粤剧博物馆使用，为满足展览需要，在青云巷内原建筑物结构之外加建了便于展示藏品的钢结构玻璃顶，使得新、老材料相得益彰。至此，整个修复工程全部完成。

回顾整个修缮过程，在研究、设计和施工工程中追求材料、工艺和构造的价值，恰当地表达出建筑材料和工艺的表现力，恪守修缮中改善结构、构造和功能的本分而不过多铺陈装饰，是一次颇为有益的实践。

（鸣谢：本文在写作过程中得到了吴庆洲教授的指导，以及佛山市文管会吴庭章先生、广东五华一建谢伟光技术员和邓吉彩工程师、广东民间工艺博物馆李继光副馆长的帮助。）

【参考文献】

[1] 华南理工大学东方建筑文化研究所.兆祥黄公祠修复工程修缮设计报告[R].

[2] 吴庆洲，冯江等.兆祥黄公祠申报联合国亚太遗产文化保护奖申请文件[Z].

[3] [清]屈大均.广东新语[M].北京：中华书局，1985.

[4] 蔡易安编著.清代广式家具[M].上海：上海书店出版社，2000.

[5] [宋]赵汝适.诸蕃志校释[M].北京：中华书局，2000.

参考文献

一、相关法规条例

[1] 国际宪章、国家法律法规、准则.

[2] 中华人民共和国文物保护法实施条例（2003）.

[3] 中华人民共和国历史文化名城名镇名村保护条例（2008）.

[4] 保护文物建筑及历史地段的国际宪章（威尼斯宪章）（1964）.

[5] 中国文物古迹保护准则 国际古迹遗址理事会中国国家委员会（2002）.

[6] 关于乡土建筑遗产的宪章 ICOMOS 1999.

二、相关地方志、著作和论文

[1] 万历贵州通志.

[2] 万历黔记.

[3] 乾隆贵州通志.

[4] 民国贵州通志.

[5] 咸丰安顺通志.

[6] 乾隆佛山忠义乡志.

[7] 民国佛山忠义乡志.

[8] [宋]赵汝适. 诸蕃志校释[M]. 北京：中华书局，2000.

[9] [清]屈大均. 广东新语[M]. 北京：中华书局，1985.

[10] 杨友维，鲍中行等. 大明屯堡第一屯——鲍家屯[M]. 成都：巴蜀书社，2008.

[11] 王瑞珠. 国外历史环境的保护与规划[M]. 台北：淑馨出版社，1993.

[12] 傅熹年. 傅熹年建筑史论文集[M]. 北京：文物出版社，1998.

[13] 阮仪三. 历史环境保护的理论与实践[M]. 上海：上海科学技术出版社，2000.

[14] 贺为才. 徽州城市村镇水系营建与管理研究[D]. 广州：华南理工大学，2006.

[15] 东南大学建筑系. 徽州文物管理所编著. 棠樾[M]. 南京：东南大学出版社，1993.

[16] 鲍世行，鲍智捷. 富有特色的屯堡古村落——贵州安顺鲍屯[M]//鲍世行. 谭徐明主编. 贵州安顺鲍屯——大明屯堡古村落. 2006.

[17] 张卫东，庞亚斌. 600年鲍屯水利探考[J]. 中国水利，2007（12）.

[18] 吴庆洲. 贵州小都江堰——安顺鲍屯水利[J]. 南方建筑，2010（4）.

[19] 谭徐明，王英华，朱云枫. 贵州鲍屯古代乡村水利工程研究[J]. 工程研究——跨学科视野中的工程，2011，3（2）.

[20] 贵州省安顺市水利水电勘测设计院. 贵州安顺市鲍屯古水利保护、恢复项目建议书[R]. 2007.

[21] 彭瑛. 文化生存范式下文化遗产的旅游开发——以贵州安顺鲍屯村为例[J]. 安顺学院学报，2011，13（6）.

[22] 魏皎，汪永平，王盈. 贵州安顺屯堡聚落防御性分析[J]. 华中建筑，2010（3）.

[23] 吴庆洲. 中国古代城市水系[J]. 华中建筑，1991（2）.

[24] 鲍世行. 鲍屯的水口园林[N]. 中国水利报，2007-4-19（8）.

[25] 尹东跃. 加强古水利工程保护意义深远[N]. 中国水利报，2007-4-19（8）.

[26] 徐好好. 前现代建筑技术一课[J]. 新建筑，2015（1）.

[27] 华南理工大学东方建筑文化研究所. 兆祥黄公祠修复工程修缮设计报告[R].

[28] 吴庆洲，冯江等. 兆祥黄公祠申报联合国亚太遗产文化保护奖申请文件[R].

[29] 蔡易安编著. 清代广式家具[M]. 上海：上海书店出版社，2000.

[30] 吴运江. UNESCO亚太文化遗产保护奖综述及我院获卓越奖评价[J]. 南方建筑，2013（1）.

[31] 1930年代的"安宫牛黄丸"说明书[N]. 佛山日报，2006-4-21.

[32] 冯江. 祖先之翼：明清广州府的开垦、聚族而居与宗族祠堂的衍变[M]. 北京：中国建筑工业出版社，2010.

[33] Jukka Jokilehto. A History of Architectural Conservation[M]. Oxford: Elsevier-Butterworth Heinemann, 1999.

[34] Robert B. Marks. Tigers, Rice, Silk, and Silt: Environment and Economy in Late Imperial South China[M]. New York: Cambridge University Press, 1998.

[35] Paolo Ceccarelli. Disomogeneita, Frammentazione, Disordine, Discrezionalita: Inconsueti Suggerimenti dall' Esperienza del Guangdong[J]. Archivio di Studi Urbani e Regionali, FrancoAngeli, Milano, 2008 (92).

后记

本书介绍了联合国教科文组织亚太文化遗产保护奖获奖的2个项目——贵州安顺鲍家屯水碾房修复工程和佛山兆祥黄公祠修复工程。

佛山兆祥黄公祠始建于清光绪三十一年（1905年），占地3000m²，是佛山城区内现存规模最大，形制最完整的祠堂建筑群。1999~2001年，我们团队进行兆祥黄公祠修复工程的勘察、测绘、修复设计的工作。修复后兆祥黄公祠由佛山市市级文物保护单位上升为广东省省级文物保护单位。佛山兆祥黄公祠修复工程项目获2005年UNESCO亚太文化遗产保护荣誉奖，成为地方文化遗产保护工程的范例。

贵州鲍家屯水碾房建于明朝初年，至今已有600多年历史。2008年，我应鲍世行教授邀请，参与鲍家屯水碾房修复工程项目，成为该项目的技术负责人。当时该项目团队中有一些不同意见，认为水碾房又破又旧，不如拆掉，另建现代的钢筋水泥的水碾房。我认为这座水碾房是明代建筑，应保持原有的结构，应用当地的工匠，用传统的材料、传统的工艺和传统的技术加以修复，保持其原真性。

团队成员都支持了我的意见，2009年按此办法将水碾房修复。该水碾房原先并非文物保护单位。后来逐级申报，成为市级、省级以及全国文物保护单位。2011年，贵州安顺鲍家屯水碾房修复工程获2011年UNESCO亚太文化遗产保护奖最高奖卓越奖，成为第三世界文化遗产保护工程的典范。2013年该项目入选国际最佳遗产保护范例俱乐部，是我国第一个进入该俱乐部的文化遗产保护工程项目。

试想，如果当时没采取正确的方法，把水碾房拆了，或用现代材料把它修走样了，修坏了，我们就毁掉了一个可以作为国家级文物保护单位，可以成为第三世界文化遗产保护工程典范的文化遗产瑰宝，这该是多么可怕的后果！事实上，我们在近30多年的快速发展中，已经出现过许多这种令人痛心的事情。本书总结这一经验教训，是希望我国的文化遗产保护工程能在正确的理念指引下持续向前发展。

文化遗产保护工程要取得成绩实非易事。为此，我由衷地感谢鲍世行先生，感谢他邀请我参加鲍家屯水碾房修复工程项目，感谢他支持了我的正确意见。我由衷地感谢该工程项目的团队成员和鲍家屯村委会，正是由于团队成员的亲密合作和村委会的大力支持，才使修复工程得以顺利完成。我由衷地感谢参加修复工程项目的老工匠，他们不辞辛苦，用水碾房原有的材料，用传统的工艺，传统的技术，修复了水碾房，使其既保持了原真性，又焕发了生机，继续为村民碾米、磨面，恢复了鲍家屯二十四景之一的“碾房听音”这一传统文化景观。

至于佛山兆祥黄公祠，我也由衷地感谢我的项目团队成员的亲密合作和团结一致，使工程勘察、测绘、修复设计得以顺利完成。我由衷地感谢该项目的施工单位广东省五华第一建筑公司，没有他们的密切配合，细致施工，保质保量，精益求精，兆祥黄公祠就不可能如同现在那样焕发光彩，更不可能成为地方文化遗产保护的范例。

我由衷地感谢冯江和徐好好，本书的两位合作者。冯江老师和徐好好老师既是我的学生，又是我的工程项目合作者。项目成果的取得和获奖，他们俩功不可没！

我由衷地感谢意大利费拉拉大学终身教授，联合国教科文组织地方性城市和区域可持续发展规划委员会主席齐珂理先生，百忙之中为本书作序，并对本书的2个工程项目的开创性贡献作出了精到的评价。

我由衷地感谢一切关心、支持和帮助过我们文化遗产保护工程的单位和个人！

文化遗产保护工程任重而道远。

“路漫漫其修远兮，吾将上下而求索”。

希望本书在同仁们的求索中可以作为有益的参考。

吴庆洲

2014年12月1日于广州